Building Stairs

Building Stairs

ANDY ENGEL

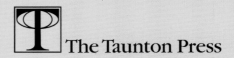

The Taunton Press

The Taunton Press
Inspiration for hands-on living®

The Taunton Press, Inc., 63 South Main Street,
PO Box 5506, Newtown, CT 06470-5506
e-mail: tp@taunton.com

Editors: Peter Chapman, Robyn Doyon Aitken
Jacket/Cover design: Alexander Isley, Inc.
Interior design: Lori Wendin
Layout: Carol Petro
Illustrator: Chuck Lockhart
Photographers: Andy Engel and Patricia Steed

For Pros/By Pros® is a trademark of The Taunton Press, Inc.,
registered in the U.S. Patent and Trademark Office.

Library of Congress Cataloging-in-Publication Data
Engel, Andy.
  Building stairs / Andy Engel.
       p. cm. -- (For pros/by pros)
  Includes index.
   ISBN 978-1-56158-892-3
 1. Stair building. 2. Finish carpentry. I. Title.

TH5670.E54 2007
694'.6--dc22

                              2007008994

Printed in China
10 9 8 7 6 5

The following manufacturers/names appearing in *Building Stairs* are trademarks:
Bosch®; Collins Tool Company®; Elmer's®; Fasten Master®; LedgerLok®;
Festool®; KregJig®; Masonite®; Osmose℠; Pl 400®; Porter-Cable®; Saran® Wrap;
Titebond®; Trex®; Trujoist®

Acknowledgments

*To my late parents,
Sheila MacArthur and
Charles "Ink" Engel.
One taught me to love books,
the other to love working
with my hands.*

Most of what's in this book I learned from others. A great deal of whatever stairbuilding and carpentry know-how I possess was gleaned from magazines such as *Fine Homebuilding* and *JLC*. Those magazine authors—professional builders and amateur writers that they are—rarely receive the credit they deserve. With the dissolution of the master/apprentice system, these authors are the ones carrying forward the collective memory and traditions of centuries of builders.

With this in mind, it would be dishonest to say that I alone wrote a book. I was clearly more of an organizer of information than a creator.

Just as words don't spring onto the page without the help of others, neither do photos. For the hours she spent learning to use my fancy new camera, and then using it so well that she frequently annoyed me by being right, I thank my wife and oldest best friend, Patricia Steed. I also thank my sons, Duncan and Kevin Engel, who were regularly pressed into service when their mom was busy.

Others deserve my thanks as well, more, I'm certain, than I've remembered. Several are: Werner Thiel, for teaching me to be a carpenter; Kevin Ireton, for teaching me to be a writer; Jed Dixon and Stan Foster, for graciously sharing their stairbuilding knowledge; Lynn Underwood, for his encyclopedic knowledge of the International Residential Code; Peter Chapman and Robyn Doyon Aitken, my editors, for doing their jobs so well; Wendi Mijal, for making my poor photos look good; and the staffs at *Fine Homebuilding, Fine Woodworking,* and *JLC*, present and past, for all they've taught me about carpentry, writing and photography, and for their friendship.

Contents

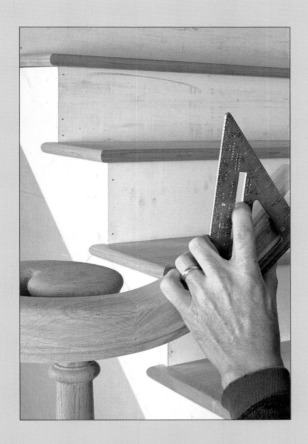

Introduction

S tairs are at once **utilitarian** and beautiful. Building them and their railings is both the height of the carpenter's craft and a mundane combination of basic carpentry and seventh-grade math. Any trim carpenter, and most owner-builders, already possess the basic woodworking tools and skills. And anybody who can add, subtract, multiply, and divide, as well as sketch a project on graph paper, has the necessary ciphering tools.

If you've read this far, most likely you've got the tools and have completed seventh grade. What's left is the question of inclination. Stairs require careful craftsmanship, but they aren't art. They're craft. David Pye, in *The Art and Nature of Workmanship*, defines *workmanship of risk* and *workmanship of certainty*. Artists undertake workmanship of risk. If Michaelangelo twitched on his final chisel stroke, he might have whacked David's nose clean off. Craftsmen rely on workmanship of certainty. When it really counts, they use aids such as jigs to make, for example, straight and square cuts. Stairbuilding is almost all workmanship of certainty.

That said, stairs do separate the carpenters from the hackers. Even the most rustic basement stairs built from framing lumber must be sturdy and consistent in tread (the parts you walk on) width and riser (the vertical space between treads) height. When a carpenter builds a wall or a roof, he can be off by a surprising amount—inches sometimes—and the home-owner might never know. Stairs aren't like that. Discrepancies as small as ¼ in. become at worst a trip hazard, and at best an annoyance. Sound exacting? It is, but by employing tools of workmanship of certainty such as a tablesaw and a rip fence, staying within such tolerances barely merits a second thought.

Moving out of the basement and into the foyer, stairs and their railings, or balustrades, can assume an almost sculptural quality as the dominant element of a home's entry. Formal stairs are as much like furniture as anything a carpenter does. But furniture makers ply their craft on a bench in a shop where perfection is possible. The stairbuilder's work must seamlessly fit inside the usually imperfect world of homebuilding. The furniture maker's fit and finish, which formal stairs seem to demand, keep a lot of carpenters from even trying to build them. Here's the thing, though—stairbuilding is just another way to apply carpentry skills. Although I usually build some types of stairs in the shop, that's out of

Even rough stairs must be cut precisely. **Inconsistency at this stage reflects in the final product, presenting a daily trip hazard.**

convenience, not necessity. I've built every type on-site as well.

Like any trade, stairbuilding can seem impossibly complex. And when learning, one's own assumptions can get in the way. Stairbuilding, more than any other aspect of carpentry I've learned, requires an open mind and an occasional leap of faith. I once had an apprentice who was stymied by the idea of running a diagonal structure across the plumb and level world of homebuilding. He just couldn't understand how to measure and plan that theoretical line in space where the stringer would rise between two floors. It rocked his world when I told him that I don't measure that diagonal. I look at the vertical rise, and from that I figure the number and height of the risers. I look at the horizontal distance the stair is to traverse and figure the width and number of treads from that. It took days before he got comfortable with the fact that what he'd assumed for years as being a key to the

Workmanship of risk, the use of unguided cutting tools, as shown in the mortising of this newel for a rail, is rare in stairbuilding.

mystery of stairbuilding was incidental and largely an impediment to his learning.

In this book, I'll explain the basics of stair geometry and planning so that you can build stairs to fit any opening you encounter. You'll learn to build the most basic stairs by notching framing lumber for the stringers and screwing down rough treads. From there, it's a small leap to routing mortises in the stringers and building stairs whose assembled parts look as if they were cast in a mold.

The second part of this book is about railings. In 1985, when I first started in stair work, railings scared me. Much like my apprentice, that diagonal line through space overwhelmed my thinking. In a way, though, railings are easier than stairs. That rail should end up parallel to the existing stair, and all

you need to do for that to happen is set its posts the same height off the stairs. And those curved parts that look as if they grew off the end of the railing? They're off-the-shelf parts that you join to the rail with a bolt. The secret to making them look seamless is sandpaper and elbow grease.

If you can measure the distance between two floors of a house and divide that number by 7 or 8, you can plan out a set of stairs. If you're competent with a router and a circular saw, you can cut a stringer. Can you read a level and operate a drill and a miter saw? You can learn to install a railing. You probably already possess most of the skills needed to build stairs. What this book will do is fill in a few gaps and show you how to apply some pretty basic carpentry in a way that can yield stunning results.

Stairbuilding requires both shop and site work. This bullnose starting step was made in a shop but joined to the stair on site.

Houses aren't ever perfectly plumb, level, square, or straight. Stairbuilding skills include fitting finish work to such imperfect conditions.

Stairbuilding Basics

For me, the fun part of building stairs is the woodworking—a chance to practice what many regard as the height of a carpenter's skills. But before touching a piece of wood or a saw, you have to plan the stairs to comply with the building code and to fit the existing space. Before you can plan the stairs, you have to understand the geometry behind stairbuilding. At its most basic, planning a stair involves figuring out how many risers (the vertical parts of a stair) and treads (the bits you walk on) there are and how big they have to be.

Safety is without question the highest priority when planning and building stairs. Poorly designed or executed stairs are flat-out dangerous. My childhood neighbors lived in an old farmhouse with a steep, uncomfortable, narrow-treaded stair. At age four or five, I tumbled all the way down those stairs. More scared than hurt, I was lucky. In most years, the Consumer Product Safety Commission estimates that there are more than a million stair-related injuries serious enough for medical attention.

Good looks are only part of the package. A quality stair has consistent, comfortable treads and risers and a sturdy and safe balustrade.

The Parts of a Stair

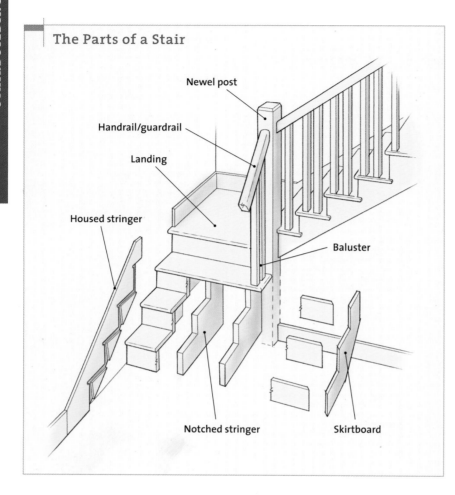

Newel post

Handrail/guardrail

Landing

Housed stringer

Baluster

Notched stringer

Skirtboard

Follow the Code

Today's building codes would not allow those old farmhouse stairs in new construction. Codes are quite specific about such things as rise, run, and width. Most states have now adopted the International Residential Code (IRC), the intention behind which was to replace the three older model codes with a uniform national building code (see Appendix B: Codes on p. 229). That was a great idea, but in practice, states and local jurisdictions remain free to modify the IRC, and few hesitate to do so. There seems to be little consensus between jurisdictions regarding stairs. For example,

while the 2006 IRC specifies a maximum rise of 7¾ in. and a minimum run of 10 in., my own state, Connecticut, currently permits a rise of 8¼ in. and a run of 9 in., the same as it was under the older Council of American Building Officials (CABO) and Building Officials Council of America (BOCA) codes. Some states have kept the IRC's 7¾-in. rise but reduced the minimum run to 9 in. The bottom line is that you need to check with your local building department to get the specifics on what constitutes a legal stair where you're building.

Stair pitch

It may seem that the difference of ½ in. in rise or 1 in. in run within the various jurisdictions' codes is inconsequential, but it's not. Shorter risers and deeper treads make a shallower-pitched stair that's easier to climb. Pay attention the next time you use the stairs in a public building. Stricter codes govern public buildings, and their stairs tend to be quite shallow. You'll notice the difference. Changing demographics are one more reason that modern codes call for shorter risers and deeper treads. In this, the day of the aging, squeaky-kneed baby boomer, shallower stairs are gaining appreciation.

The other side of the coin shouldn't be ignored, either. Using a rise of 8¼ in., when legal, usually allows you to sneak in a 13-riser stair into a house with 8-ft. walls and 2x10 floor joists, the most common configuration of house that I've worked in. If your stairs will be installed where the code allows only a 7¾-in. rise, then you'll have to build the stairs with 14 risers. What difference would it make to add a riser? Adding a riser adds a tread, which requires at least another 9 in. or 10 in. of space on both floors. That may not sound like much,

Rise, Run, and Headroom

Building codes are specific regarding stair geometry, but they do vary regionally. Dimensions referenced here are from the 2006 International Residential Code (IRC). *Rise* is the vertical distance traveled with each step. It is calculated by dividing the floor-to-floor height (total rise) by the number of steps. Maximum rise is limited by the building code.

Run, or unit run, is the horizontal distance covered with each step. It's the distance between the faces of successive risers, not the width of the tread. Tread width is the run plus the amount that the front of the tread, or *nosing,* overhangs the face of the riser below. In a typical stair, the run might be 10 in. and the nosings 1¼ in., making the tread width 11¼ in.

Headroom is measured plumb up from the line of the tread nosings. In the IRC, the minimum headroom is 80 in.

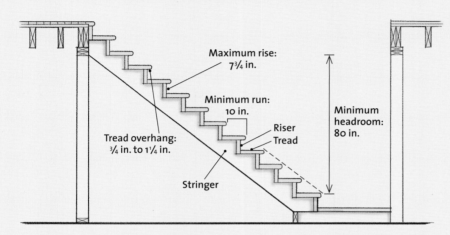

Maximum rise: 7¾ in.

Minimum run: 10 in.

Tread overhang: ¾ in. to 1¼ in.

Riser
Tread

Stringer

Minimum headroom: 80 in.

Finding the Riser Height and the Total Run

To find out how long a stair will be, its total run, you first have to determine how many treads there will be and what the run of each tread is. There is always one fewer tread than there are risers, so the first step is figuring out the number of risers. Minimum run is specified by building codes, and it's often 9 in. or 10 in. The run can be larger, but the rule of thumb is that the sum of the rise and run should be between 17 in. and 18 in.

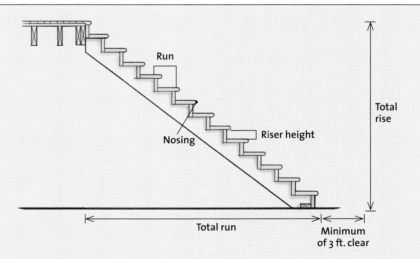

Run

Nosing

Riser height

Total rise

Total run

Minimum of 3 ft. clear

Riser height = total rise ÷ number of risers

Total run = run x number of treads + thickness of top riser + bottom nosing

Total rise of 107⅞ in. ÷ 14 risers = 7⁴⁵/₆₄ in., which rounds to a riser height of 7¹¹/₁₆ in.

10-in. run x 13 treads = 130 in. Add ¾ in. for the top riser and 1¼ in. for the bottom nosing for a total run of 132 in.

Tread depth = run + nosing

The top riser fits behind the top tread.

but it's a lot in a small house. Nine inches might crowd a door in a foyer, complicate (and increase the expense of) floor framing where a basement stair enters a kitchen, or steal a third of the depth of a closet.

Keep it consistent

Another specific part of the IRC regards consistency in riser height and tread width. Only a ⅜-in. deviation from the greatest to the smallest is allowed and only a ³⁄₁₆-in. deviation between consecutive treads or risers. In fact, consistency is more important than tread and riser size because we quickly adapt to regularity, even if it's not terribly convenient.

A ladder is a stair whose treads and risers meet no residential code. Yet we've all gone safely up and down them, at least in part because we quickly learn their consistent rhythm. Achieving that level of consistency isn't hard, if you pay attention.

There's a lot of advice out there on building comfortable stairs. Some people say that the unit rise should fall between 7 in. and 7½ in. and that the unit run should be between 10 in. and 11 in. You'll find ratios suggesting that the sum of one riser and one tread should be between 17 in. and 18 in. or that the sum of two risers and one tread should be between 24 in. and 25 in. This is great in theory, but as a practical matter, stairbuilders usually have to fit their work into an existing space that may or may not allow such niceties.

As a stairbuilder, I'm typically invited into a project when the framing is nearing completion. I measure the overall distance, or rise, between the floors, as well as the length and width of the opening. After factoring in the thickness of the finish flooring (more on this

Nonstandard Stairs

My early tumble down a neighbor's stairs notwithstanding, I've used, and built, comfortable stairs that weren't code compliant. They worked for two reasons. First, they were uniform in riser height and tread width. That's critical with any stair because your feet quickly find the stair's rhythm without much conscious thought. Deviate from that rhythm, and you're off balance. Second, there's a proportion between rise and run that most stairs fall into. Stick with that, and we tend to find our feet. There's an old rule of thumb that the sum of the riser height and the tread width should fall between 17 in. and 18 in. I've used that rule to fit stairs in older houses. In remodeling, it's not uncommon to encounter an existing stair opening that doesn't allow for a legal set of stairs without extensive reframing. In those cases, I've had excellent results by approaching the local building inspector and explaining the problem. Inspectors have broad, de facto authority to approve deviations from the code. In such older houses, the new stairs I built were always an improvement over the existing stairs, and I always kept the sum of the rise and run between 17 in. and 18 in.

These stairs may once have been safe, but the crazy quilt of varying riser heights and out-of-level treads is an accident waiting to happen.

Stair width is measured between the upper-floor framing. **Be sure to measure the side of the opening where the stair reaches that floor.**

Working On-Site or in the Shop?

I've built stairs both on-site and in my shop. Now, my shop is no great shakes, but it's where I prefer to do my stairbuilding. The chief advantage is ready access to two stationary power tools: a tablesaw and a jointer. Even though my tablesaw is old and has significant quirks, it's far superior to even the best job-site saw. And the jointer's ability to straighten a board and quickly remove saw marks from a ripped edge are advantages that you have to experience to appreciate. Plus, lunch and an afternoon latte are only steps away.

Another advantage of working in my shop is that it's heated. That's not just for my comfort. Lumber and glue fare far better when stored and used in warm environments.

The chief disadvantage to working in the shop is transporting the stairs. A full set of oak stairs weighs several hundred pounds and takes up a lot of room. I get them out of the shop as soon as possible. Stairs up to 14 risers will stay in the back of my 8-ft.-bed pickup, tied in and counterbalanced with something like a shorter set of stairs or some heavy tools. Longer stairs have to go on a trailer.

Working on-site provides many second chances for checking measurements and test-fitting stringers. But job sites can be cold, miserable places, crowded with other tradesmen competing for the same space. All told, I'll take my shop.

later), I can quickly figure out the rise and run of the stair. It's often tempting to add a riser and a tread to build a more comfortable stair with a shallower rise. Be careful. Code also specifies a minimum amount of headroom, typically 80 in. measured plumb from a line tangent to the tread nosings (see the top drawing on p. 9). If the house's designer planned a steeper stair than you're thinking of, adding that extra tread sometimes raises the line of the stair enough to create a headroom problem.

Stairwell width

Finally, building codes specify minimum stairwell widths. Here, codes are usually consistent. Above the minimum railing height (generally 36 in. above the line of the tread nosings), there must be at least 36 in. between the finished walls of an enclosed stairway. This allows room for the thickness of the stringers, and some codes permit stairways to attics to be as narrow as 32 in. In practice, I like stairways to be as wide as possible to make it easy to move

furniture or pass your housemates. If I've got a say in the house's design, I suggest stairs that are 42 in. or more wide.

Measuring for Stairs

About 15 years ago, I measured incorrectly for a set of housed-stringer basement stairs. I've long forgotten how I managed this feat. I probably stopped at the job on my way home from a day's work, tired and harried. The why doesn't matter, except as a caution against doing anything important when you're not at full capacity. I took the measurements back

Measuring the distance between floors

1 Get the overall height. Have a helper stretch the tape measure to ensure an accurate measurement of the overall rise. To avoid math, if there are samples of the finished floor material available, it's a good idea to set them in place when measuring.

2 Check for level. If one or the other floor isn't level, where you measure the overall rise becomes important. Pull the tape plumb from where the stair will land on the out-of-level floor to the other floor.

3 If both floors are out of level, you'll need to create a level reference line between them. Measure to it from the top and bottom landings, and add those heights to find the overall rise.

to my shop and confidently built the stairs. I didn't discover that they were too short until I tried to install them. I dropped them into place in the cellar and immediately saw that the treads sloped forward at a pitch that would easily drain water. At that moment, I realized that a day's work and a couple of hundred dollars' worth of perfectly innocent poplar was now little more than expensive firewood.

It's absolutely critical to measure three things: the overall distance between the floors, the width of the stairwell, and the length of the stairwell. I always sketch the stairwell to keep a reference for thinking the stair through.

Measuring the distance between floors

Measuring the overall distance between the floors is usually a straightforward matter of stretching a tape measure. The best approach is to have a helper hold the bottom of the tape tight to the lower floor while you read off where the tape hits the top of the upper floor, but I've done it alone plenty of times. Simple enough, but there are a couple of catches.

You have to be sure the floors are level or at least parallel with each other. Check parallel by measuring the difference between the floors at both the head and the foot of the stairs. If the measurements are equal, the floors are parallel. In new construction, floors are usually level enough. Most of the out-of-level floors I see in new houses are concrete-slab floors, which in the Northeast typically means the basement. Where slabs are common on the first floor, you'll encounter this problem with the main stairs.

If the upper floor is level, there's no problem. Simply measure straight up from about where the stairs will land on the lower floor. If the problem is reversed and the upper floor

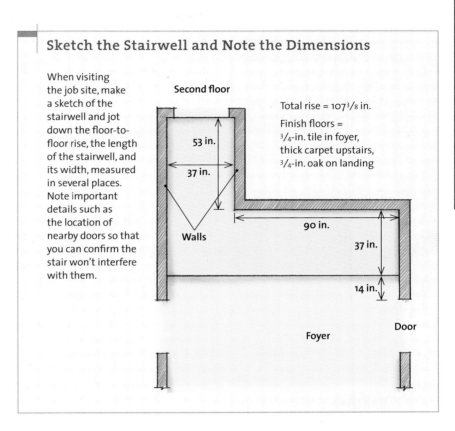

Sketch the Stairwell and Note the Dimensions

When visiting the job site, make a sketch of the stairwell and jot down the floor-to-floor rise, the length of the stairwell, and its width, measured in several places. Note important details such as the location of nearby doors so that you can confirm the stair won't interfere with them.

Second floor

Total rise = 107 3/8 in.

Finish floors = 3/4-in. tile in foyer, thick carpet upstairs, 3/4-in. oak on landing

53 in.

37 in.

Walls

90 in.

37 in.

14 in.

Foyer

Door

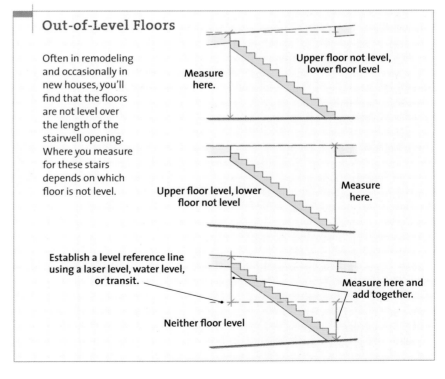

Out-of-Level Floors

Often in remodeling and occasionally in new houses, you'll find that the floors are not level over the length of the stairwell opening. Where you measure for these stairs depends on which floor is not level.

Measure here.

Upper floor not level, lower floor level

Upper floor level, lower floor not level

Measure here.

Establish a level reference line using a laser level, water level, or transit.

Neither floor level

Measure here and add together.

Out-of-Level Landing

When the floor perpendicular to the run of the stair is out of level by more than 1/8 in. over 3 ft., the affected riser has to taper. Measuring the overall height at the stair's centerline and splitting the taper between the sides maintains the most consistent riser height.

Out-of-level landing

Add this distance to the top riser height, scribe, and trim to fit.

Floor framing

CL

Measure to the top of the finished floor at the stair's centerline.

Stringer location

Lengthen the bottom riser cut in the right stringer by this amount.

Shorten the bottom riser cut in the left stringer by this amount.

Taper the bottom riser to fit the stringers.

As a professional stairbuilder, I ask my customers to sign a paper describing the finished floor choices. That way, if they decide to change the flooring after the stairs are in, I've got evidence that the stairs were built to specifications.

is out of level, then you measure to where the stairs reach that floor. If both floors are out of level and out of parallel, I take a trick from North Road Stair's Jed Dixon and establish a level reference line using a long level (a laser level would be better or a water level would work, too). Measuring to the level line from the stair's upper and lower landing points, then adding these numbers together yields the stair's overall rise.

If either floor is out of level parallel to the run of the stairs, it's not a big deal. If a floor is out of level perpendicular to the run of the stair, then some adjustment needs to be made. For the stair treads to be level, the riser that hits the out-of-level floor, be it upper, lower, or both, has to taper. I measure overall rise in the center of the stair, where people are most likely to walk. Then I split the difference, with one side of the riser being slightly shorter and one side slightly taller.

In a perfect world, all measurements for stairs would be taken after the finished floors are in place. Of course, stair measurements are typically taken during the framing stages. It's crucial to learn the thickness of the finished flooring materials that will be added to the subfloor because this thickness affects the overall rise of the stair. Assuming that you're measuring from subfloor to subfloor, you add the thickness of the upper finished floor to the overall rise and deduct the thickness of the lower finished floor. If, for example, there will be 3/4-in. hardwood flooring on the lower floor, deduct 3/4 in. from the overall height of the stair. If there's to be 3/4-in. hardwood on the upper floor, add 3/4 in. to the overall height.

The finish material on the first floor is typically straightforward. Most main stairs end up in a foyer, which is usually finished with a rigid material—tile, stone, or hardwood. The same goes for basement stairs. The lower end lands on concrete, and often the upper end is in the kitchen, which normally receives tile, hardwood, or vinyl. My bugaboo is where the main stair reaches the second floor. If the second floor is 3/4-in. hardwood that matches the stock landing tread I can buy in any lumberyard, that's easy. It's when the second floor is to be carpeted that I need to make some choices.

Accommodating Finish Flooring at the Top of the Stairs

Depending on the floor treatment, the landing tread is installed flush with the subfloor or on top of it. In drawing 1, stock landing tread matches the hardwood flooring's thickness, so it is installed on top of the subfloor. Drawing 2 shows a stair that's to be wrapped with the same carpet as the upper floor, so a strip of subfloor is removed and the landing tread installed so its top is even with the subfloor. In drawing 3, the second floor's carpet is thick enough to butt to the landing tread installed atop the subfloor. Easing the landing tread's edge makes it friendlier to bare feet. Drawing 4 shows a thin carpet, such as Berber, terminating just on the landing tread installed even with the subfloor. Engineered-wood flooring as shown in drawing 5 is typically thin—3/8 in. or thereabouts. In that case, the stock landing tread's rabbet must be deepened so the remaining material matches the thickness of the flooring.

In all cases, the landing tread's overhang must equal that of the stair treads. When calculating the stairs, it may help to remember that the landing tread's top is the top of the overall rise, no matter its relation to the subfloor. If there is a newel, I don't install the landing tread until it's time to do the railings. It's easier to fit the landing tread around the newels than vice versa.

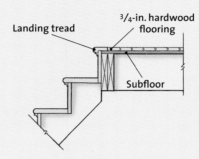

1. Top riser even with subfloor

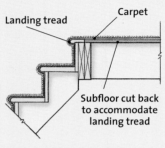

2. Top riser even with framing

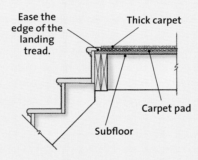

3. Top riser even with subfloor

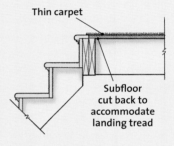

4. Top riser even with framing

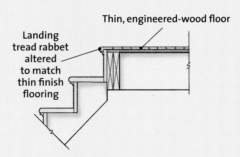

5. Top riser even with subfloor

Landing tread transitions between the stair and the upper landings. **Stock landing tread is made from 5/4 material, as are most stair treads. It is usually 3¹/₂ in. wide and rabbetted in the back to a ³/₄-in. thickness, which matches nicely with traditional hardwood floors. The leading edge overhangs the top riser, just as a tread would.**

Look for a bug. **If a window is within 36 in. of a stair or landing (18 in. if the glass is separated by a code-approved railing) or within 60 in. of the bottom tread and less than 60 in. higher than the tread nosing, it must be glazed with safety glass. The most common is tempered glass, which you can spot by looking for the "bug."**

The next thing to measure is the width of the stairwell. The distance between the framing had better be at least 37 in. to allow for ½-in. drywall on each side and still meet the minimum required width. If it's narrower than that, somebody needs to have a talk with the framing carpenter.

Finally, note the length of the stairwell, particularly where any turns have to go, and sketch a plan view of the whole kit and caboodle. It's also a good idea to note the location of any doors so that you can check that the stair won't interfere with them. In addition, check the location and height of any windows. With this information complete and in hand, you can start planning the stair.

Planning a Stair

The most basic stair has a straight run—that is, there's enough space for it to go from floor to floor without turning at a landing. It's fairly common to encounter straight stairs that descend into basements, and sometimes main stairs are configured this way. Most stairs I build, though, incorporate a landing and a turn. There's good reason why architects design stairwells in this way. Landings add a safety factor. In fact, the IRC requires a landing in any stair that exceeds an overall rise of 12 ft. On long stairs, this is to provide a break from climbing. On any stair, a landing minimizes the distance the clumsy might tumble before stopping. Landings must be at least as wide as the stair and at least 36 in. in the direction of travel.

Basic Stair Configurations

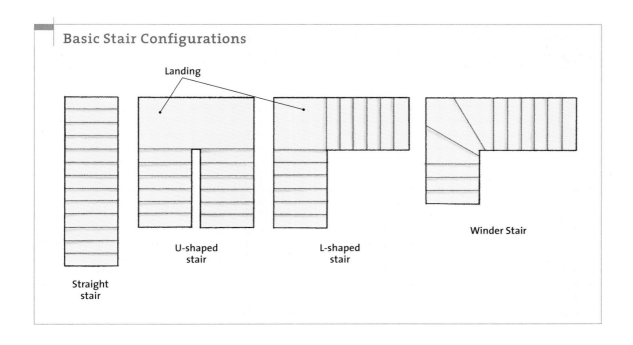

Landing

Straight
stair

U-shaped
stair

L-shaped
stair

Winder Stair

Straight run or turn?

Determining whether a stair can be a straight run or if it requires a landing is a simple matter. Odds are that all you have to do is look at the framing. If the stairwell turns, there's probably a landing (or winders) at the midway point. Sometimes not, though. Many straight stairwells are designed for a landing at the very bottom—one that's only one or two risers above the first floor.

To noodle this out, first measure the overall rise, factoring in the finish flooring. Divide this number by the maximum allowable height, and round up to find the number of risers. Under the IRC, that's typically 14 for houses with 8-ft. ceilings. Nine-foot ceilings usually require 16 risers. (Remember, it's not just wall height that matters. You must also include the thickness of the upper floor framing.) Dividing the overall rise by the number of risers yields the individual riser heights. Make sure this number is within your local code parameters.

Stairs turn at landings, which are built during the framing stages. Stout construction is called for, as a landing supports both flights of stairs.

Finding Landing Heights

Typically, the framing carpenter won't build stair landings until the stairbuilder provides the dimensions and heights. Lots of times, this task is left completely to the stairbuilder. Before the height of a landing can be calculated, its location must be laid out. And don't forget the finish flooring. Landings usually have the same finish flooring as one floor or the other.

Often, a long run of stairs will end in a landing that's one or two risers above the lower floor. If there's a wall at the bottom of the stairs, you have to figure the total run of the stairs to be certain that the landing will be at least 36 in. deep in the direction of travel. Because I'm measuring to the framing at this point, I add 1 in. to allow for drywall and a slight fudge factor.

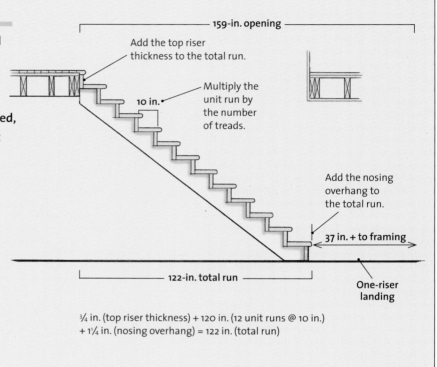

159-in. opening

Add the top riser thickness to the total run.

10 in.

Multiply the unit run by the number of treads.

Add the nosing overhang to the total run.

37 in. + to framing

122-in. total run

One-riser landing

¾ in. (top riser thickness) + 120 in. (12 unit runs @ 10 in.) + 1¼ in. (nosing overhang) = 122 in. (total run)

Blueprints and Stairs

A word of caution here. Architects and designers consider stair geometry when planning the house. Some specify the riser height, tread depth, and stair width on the blueprints. Never trust such numbers. It's not so much that the plans are likely to be wrong—although that's not unheard of—but that the house may not have been framed exactly to the plans. Maybe a wall was moved, changing the size of the opening. Perhaps someone decided to upgrade the floor joists, increasing the overall height. Maybe the carpenters framed the stairwell wider or narrower than on the plans. No matter what happened, as the stairbuilder it's your job to make stairs that work.

If your maximum code-allowed riser height is 8¼ in., you might get by with 13 or 15 risers, depending on how the finish-floor materials affect the overall rise and on the size of the upper-floor joists. Now, just because you can doesn't mean you should. If there's enough room, I always go with the shallower risers to improve comfort and safety.

Once you know how many risers your stair needs, you can work out its length, or overall run. There is always one more riser than there are treads. So deducting one from the number of risers tells you the number of treads. For example, a 14-riser stair requires 13 treads. Assuming that you're using the IRC minimum run of 10 in., then 10 in. multiplied by 13 treads yields an overall run of 130 in. Add the thickness of the top riser and the overhang of the bottom tread, which usually increase the overall run by about 2 in. The total run of that stair would be 132 in.

When you've calculated the overall run, compare it with the framed stairwell. Does it land on a floor, as opposed to somewhere in the well hole for the basement stairs? If it lands short, then either the well below is too long or you have to increase the run of the stair by using deeper treads. Is there at least 3 ft. of clear floor beyond the bottom riser? If there is less than 3 ft. of floor space, you'll need to add a landing and turn the stair.

Stairs often turn at a landing that's about halfway between the floors, a good place to take a break. It also provides a chance to show off some fancy railing work, but long before that happens, somebody has to decide the landing height. Sometimes this is specified by the architect, but I never trust blueprints for this sort of information. Too many people have had the chance to change some dimension or another, and for that matter, framing lumber itself can vary from the architect's specifications. To avoid problems, I always make a point to calculate the landing's height and give that information to the framing carpenter. Or, often, I end up framing the landing myself.

That's the theory, anyway. The next step is to take it, mix in some tools and some lumber, and build some stairs. Don't worry if you haven't wrapped your head around every detail here. Pay close attention to the code stuff, and when you get to laying out stairs, go slow. Double- and triple-check yourself. If something seems wrong, it probably is. Don't proceed until you satisfy that nagging voice in the back of your head. Conversely, I find that when things seem to be going really smoothly, I've forgotten something important. So, think things through, draw them out on paper or scrap lumber if need be, and never hurry.

Landings That Turn

In cases where the stairwell turns, the landing location is pretty straightforward. To find landing height, you must first calculate the number and height of the risers. Then, you figure out how many treads will fit between the upper floor and the turn in the framing where the landing will be. In the plan view of the example, the portion of the stairwell leading from this turn to the upper floor is 43 in. That will allow four 10-in. unit runs, with room to spare for the thickness of the top riser and for the bottom tread's nosing.

Knowing there are four units of run, you also know that there are five risers. So, the height of the landing's subfloor is five riser heights (plus the thickness of the landing's finish floor) below the upper floor. You can also figure this from the lower floor in a similar manner, the main difference being that you lower the height of the landing by the thickness of its finished floor.

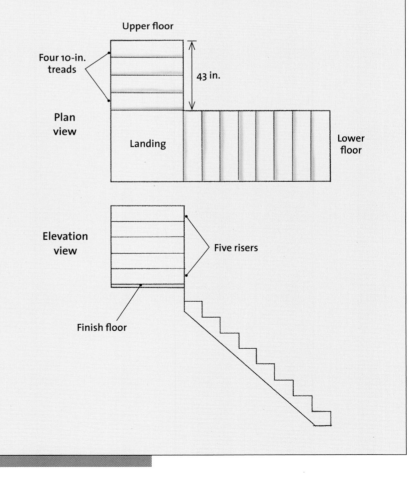

29 | Laying Out the Stringers

33 | Installing the Stringers

35 | Installing Treads and Risers

38 | Building the Support Wall

Building a Basic Stair

When a kindergartener draws a set of stairs, they're most likely notched-stringer stairs. Notched stringers, where triangular pieces are removed from the stringer stock to accommodate and support the treads and risers, are the most intuitive way to turn the theory of stairs into something you can climb. Most carpenters have built at least a few sets of stairs with notched stringers. It's a quick approach and requires tools no more specialized than a framing square and a circular saw.

For a utility stair, say, one going to a back porch or to a basement, notched-stringer stairs are a great option. Utility versions are fast to build and take tools that most carpenters have in their trucks. In many parts of the United States, particularly the West and South, the main stairs of most houses are built this way, too. They start out as rough stairs, then later on the finish carpenter adds the fancy treads, risers, and finish stringers (or skirtboards). In the Northeast, where I've always worked, most interior stairs are of the housed-stringer variety. I'll discuss the advantages of housed stringers in chapter 4.

Notched stringers are the most intuitive type of stairs. **Looking at the stringers in place, even a child can see that all that's needed to make them useful is a set of treads.**

The Framing Square

If carpentry has a sacred icon, this is it. It's the tool we use to convert theoretical values for rise and run into a stair layout on a piece of stringer stock (or into the plumb and level cuts of a rafter, but that's a different book). Understanding the use of a framing square is a gateway skill for stairbuilders. Without that understanding, you aren't building stairs.

A traditional carpenter's framing square has two legs. One, called the blade, is 2 in. wide and 24 in. long. The other, called the tongue, is 1½ in. wide and 16 in. long. The edges of the

The Framing Square Up Close

On traditional framing squares, five of the eight edges are marked in familiar eighths or sixteenths. These are the edges you'll want to use. The other three edges are marked in tenths and twelfths. Confusing these scales can lead to some interesting layout errors, and you may be wondering just what is the point of these oddball increments.

Back in the day before construction calculators gave answers in eighths and sixteenths, long division yielded answers in tenths and hundredths (it still does). The tenth scale allowed a carpenter to easily translate decimals to sixteenths. Some squares, like the one shown, have a scale for converting hundredths to sixteenths. This feature can still come in handy. Similarly, the twelfth scales are mainly used to convert tenths and sixteenths to twelfths, useful in converting long-division results into inches and in scaling conversions.

Other tables found on framing squares are used to determine rafter lengths for all sorts of roofs. The Essex Board Measure is used to quickly calculate board feet. The brace tables provide the hypotenuse for right triangles of various leg configurations, useful for cutting timber braces.

Not all square edges are the same. This side of a traditional framing square has its edges marked in eighths and sixteenths; it is the side most commonly used.

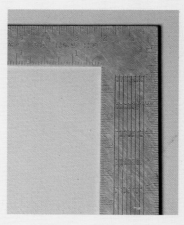

Watch out for this side. Only one scale is in sixteenths (the inside of the blade). The other scales are tenths (inside of the tongue) and twelfths. The markings can be used like a slide rule for calculations, but they confuse the heck out of stair layout.

Using a framing square for layout

1 With the stringer's crown facing you, start the layout by aligning the rise dimension on the outside of the tongue and the run dimension on the outside of the blade with the edge of the stringer.

2 Stair gauges minimize operator error, and save a fair amount of time. Tighten the thumbscrews to clamp the gauges to the square, and they make repeating the rise and run almost a brainless operation.

3 Mark the first tread and riser by running a pencil along the square. Rotating the pencil between your fingers while drawing the line keeps it sharp.

4 Slide the square along the stringer until it aligns with the pencil mark you just made. Mark this tread and riser, being careful to align these marks exactly with the intersection of the previous tread or riser and the stringer edge, and repeat down to the bottom riser.

5 To mark the level cut, where the stringer rests on the floor, align the square on the bottom riser with the tongue running back. Make sure to deduct the thickness of one tread from the bottom riser height.

The framing square is the most basic stair tool. **Understanding how to use the framing square to lay out stringers is a gateway skill for stairbuilders.**

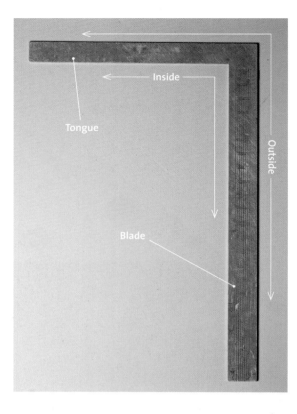

Tongue

Inside

Outside

Blade

Build On-Site or in the Shop?

There are good reasons to use site-built main stairs with notched stringers. The framing carpenter is generally responsible for cutting and installing the stringers and rough treads and risers (typically of ¾-in. subflooring material but sometimes of 2x that is discarded when the finish materials are applied). The chief advantage here is that the stairs go in during framing. As housed-stringer stairs are the finished product, most builders don't install them until the roof is complete. Having rough stairs to work from increases the efficiency of the framer and of all the mechanical trades that might work in the house before the roofer eventually shows up.

Another advantage is that, since the stairs are just rough work at this stage, the clamor and chaos of construction can go on around and on notched-stringer stairs with no fear of damaging the finished product. Finally, there are times when it's just easier to build the stair on-site. For example, shop-building stairs for an older, out-of-plumb and out-of-level home leaves you with no second chances. You build the unit and bring it to the site, hoping all the while that you didn't miss anything important in the measurements. Site-built notched-stringer stairs let you finesse the fit one piece at a time.

The downsides to building finished stairs on-site are several. First, it takes more time. You build the rough stairs, and then you come back to finish them. When I build stairs in the shop, they're done. I can mass-produce all the parts, everything is square and even, and the treads and risers are housed in mortises that hide any little sins. This segues into my second issue with site-built finished stairs: Because mortises aren't used when finishing them, there's more

square are marked with inches and fractions of inches, and various tables of nearly mythic value are inscribed down the middle of all sides of the square. While fascinating to the tool junkie and useful to the carpenter, stairbuilders can ignore the tables and focus on the ruler markings (see the sidebar on p. 22).

Stair layout is an elegant procedure that just clicks with most people the first time they see it done. In brief, you lay the framing square on the stringer stock (see the sidebar on p. 23). The measurement for the run on one edge of the square and the measurement for the rise on the perpendicular edge are aligned with the edge of the stringer stock. With the square held firmly in place, you mark along its edges, making the cut lines for one riser and one tread, and repeat as needed. Stair gauges, which are stops that clamp to the edge of the square, aid with this repetitive layout. Laying out the entire stair is a little more complicated than that but not much.

of a tendency for joints to open up seasonally. Finally, site-built finished stairs use a lot more material.

Speaking of material, notched stringers are traditionally made from 2x12 framing lumber. That's really a bad choice of material for interior stairs. Engineered lumber, laminated veneer lumber (LVL) stock in particular, is a far better choice (see the sidebar below). It's a tougher call outside, as rot-resistant engineered lumber is harder to come by.

Building stairs on-site speeds other work. Because rough stairs go in during framing, all the tradesmen from the carpenters through the mechanical contractors have easier access between floors.

Building stairs in the shop is more convenient. All the tools are at hand, and the environment is controlled. And shop-built stairs take about half the time as site-built ones.

Using Engineered Lumber for Stringers

Solid lumber worked for thousands of years as stair stringers, but we've pretty much used up, or made inaccessible, the best trees. Old-growth trees have tight grain and few knots. These characteristics yield wood that's stronger and more stable than today's farm-raised trees that spring to height in the full sun.

Traditionally, notched stringers were made from 2x12s of fir, hemlock, spruce, or southern pine. Each piece of structural lumber is graded at the mill depending on its characteristics. Knots and other defects in the center of the plank have little effect on its strength, while the same defect at its edge could mean that it's useless as a structural member. You can't rip a 2x12 into two structural 2x6s. Those defects that once existed harmlessly in the center of the 2x12 are now in critical places on the 2x6. A typical stair notch takes more than half of the depth from a 2x12, leaving, in effect, a 2x6. Notching a 2x12 has the same effect on a plank's strength as ripping it.

Additionally, wood shrinkage is a problem with solid lumber. Wood shrinks to a far greater degree across its width than along its length. If the stringer shrinks after it's notched, you end up with out-of-square notches.

Because of the problems with sawn lumber as stringers, I prefer to use engineered lumber: laminated veneer lumber (LVL) or laminated strand lumber (LSL). Trusjoist® makes 1 1/4 -in. LSL stair stringer material (and provides span tables on its website: http://www.trusjoist.com).

A final advantage of engineered-lumber stringers has to do with the depth of material available. While most stairs have supporting walls, designers sometimes want a freestanding stair. In these cases, I recommend using deeper, say 14-in. or 16-in., engineered lumber for the stringers. Solid sawn lumber generally isn't available in these depths. And many engineered-lumber manufacturers will support the use of their products by having a company engineer review the design for free.

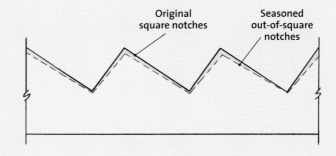

Original square notches

Seasoned out-of-square notches

Laying Out the Stair

Knowing how to lay out and notch stringers is basic to any stairbuilder. Because the layout is the simplest of any kind of stair, laying out notched stringers is the best way to learn. Mastering notched-stringer layout helps you to wrap your head around stair geometry and is a building block to more advanced techniques.

The stair shown in this chapter is a pretty big one: The opening is about 40 in. wide, and it's a 15-riser stair with a landing. You'll see a concrete ledge below the outer wall framing because this stair is in a garage—but that location doesn't materially change anything about building the stairs.

Finding the landing starting points

To get started, you have to measure the opening and go through all of the planning gyrations discussed in chapter 1. After calculating the unit rise and run and figuring out how many steps up to the landing, adding together the runs of the steps above the landing (along with an allowance for the hangerboard, a piece of ¾-in. plywood that the tops of the stringers hang from) gives the horizontal distance from the end of the stairwell to the landing. Trouble is, those points are separated vertically by 8 ft. or so.

Far faster than using a weight and a string, a laser plumb bob allows you to quickly transfer the location of the head of the stairwell to the floor. Align the upper red laser dot on the corner of the floor framing (left), and the lower dot is the corresponding plumb point below (right).

To be absolutely clear, the stairs as built in this chapter do not satisfy the International Residential Code. However, they will with the addition of finish treads that overhang the risers below. That's covered in the next chapter. And they're a great improvement over ladders placed above an open stairwell for construction access.

The solution is to drop a plumb line down from the inside corner of the stairwell, where the top of the stairs meets the upper floor. You can use a variety of tools such as a plumb bob or a level and a long straightedge. Luddite that I am in some ways, this is one place where I embrace technology. I use a laser plumb bob, which in terms of time saved on stairbuilding and remodeling jobs may be the best hundred bucks I've ever spent.

This point on the floor marks one end of the wall that's usually built under a stair; it's also the point from which you measure out to mark the landing's location (see the top photo on the facing page). I like to mark not only the end of the stringer (which in practical terms is the face of the bottom riser cut) on the floor but also the width of the bottom riser and the point 2 in. or so back where I'll start the landing. I also label these marks, spending perhaps a minute to eliminate the chance of later head scratching.

At the stud nearest to the end of the upper flight of stairs, I pencil on the landing height as measured from the lower floor (see the middle photo on the facing page). This height is simply the combined unit rise to the top of the landing and represents the top of the landing's finished floor. From there, I measure down the thickness of the finished floor and mark that

Locating the bottom end of the stair. **The location of the end of the stringer is the sum of the flight's unit runs, plus the thickness of the hangerboard, from the end of the stairwell. It's measured from the plumb point, and the other locations shown are all established from it.**

Measure up, then down. **Landing elevation is measured from the lower floor. The finish-floor height is the sum of the unit rises of the lower flight, and the height of the joists is found by deducting the thickness of the finish floor and the subfloor from that.**

Frame to a reference line. **Strike a level line from the point representing the bottom of the subfloor. Use a long, accurate level for this.**

> To check a level for accuracy, lay it on a fairly level (or plumb, depending on which vial you're checking) surface. Note where the bubble falls within the vial, then turn the level end for end. The bubble should read exactly as it did before. If it doesn't, it's trash (unless the level has adjustable vials or a lifetime guarantee).

point, then measure down the thickness of the subfloor and mark that point. This latter point is also the top of the landing's joists, and I'll strike a level line along all the existing walls at this point (see the bottom photo).

Here, I'm back to my Luddite tendencies. I like a high-quality, accurate spirit level for this sort of work. The level should be at least as long as the landing to avoid having to set the level twice. Each time you set a level, you're inviting some inaccuracy to the process. A laser level or a water level would also work here

Framing the Landing

There are as many ways to frame landings as there are carpenters. The approach I favor avoids the need for metal connectors, which rarely seem to be at hand when I need them. You can double up the outer member and use metal framing connectors, but I don't find it to be necessary. (Some jurisdictions require metal connectors, so it's always best to check.) Carpenters tend to think in terms of headers, which are usually made of doubled material. They don't always have to be, though. Headers are often doubled as part of a one-size-fits-all approach to building, done as much to save framers and architects from having to think as for reasons of strength. Headers that require doubling are designed to bear loads imposed from a much larger area, which might include

Framing the landing

1 Cut a pair of rim joists to equal the length of the landing, and nail 2x4 ledges to their bottoms. Lay the rim joists side by side so you can mark the 16-in. layout for the other joists on both rims at once.

2 Nail the inner rim joist to the wall and set the first infill joist in place. Drive two 16d common nails into each stud.

3 Run a level from the first rim joist's ledge to where a post is needed to support the outer rim joist. Deduct the thickness of the outer rim joist's ledge (1½ in.) from this number to find the post height.

4 With the outer rim supported, run the infill joists on 16-in. centers, and secure with two 16d nails into their end grain and one up through the ledger into the joist's bottom. When installed, the plywood hangerboard for the lower flight will brace the support post.

5 Don't forget the wall-side post. Since the rim joist supports all of the other joists, nails into the end grain of the wall-side joist aren't adequate support. A post must be installed here, as well.

6 More than just a walking surface, subflooring is a structural element that ties the joists together and strengthens the assembly. Secure it to the joists with adhesive and 2-in. screws.

loads from multiple floors and a roof. This is a landing that's perhaps 12 sq. ft. in size. At a design live load of 40 lb. per square foot, that's only 480 lb. (or a little more than of two of me at my winter fullest), spread evenly on the landing. Note that I don't recommend this type of framing for other floors or decks with larger tributary loading areas.

If I'm buying lumber just for the landing, 2x6s are generally more than adequate. Often, there's sufficient leftover framing lumber lying around the job, and I don't hesitate to use it up. Anything 2x6 or larger is fair game, as long as it's sound and there's enough to do the job with one size. I do avoid lumber that's been used for concrete forms but only because it wreaks havoc on a sawblade.

Setting the rim joists

One thing that slows down framing, as well as contributes to inaccuracy, is locating the members before nailing them. I avoid this on landings by nailing 2x4 ledges to the bottom of the rim joists with 16d common nails every foot or so. Not only do the ledges help to support the joists, but also the ledges automatically align the joists with the top of the rim joists.

The other joists fit between these rim joists. Nail the first rim joist to the studs of the existing wall, with its top aligned on the level line. Nail the first of the infill joists to the perpendicular wall, also along the level line, using two 16d commons per stud.

The second rim joist, the one the lower flight of stairs will abut, must be supported on both ends with posts. The height of the post on the side by the wall is easy to find—I actually measure and install it after I've nailed that rim joist to the infill joist that's secured to the wall. The other corner is slightly trickier, but

only because floors, particularly concrete slab floors, are never perfectly level. Measuring up to a level whose other end rests on the first rim joist's ledge provides the distance between the floor and the joists. Deducting 1½ in. from that for the outer rim joist's ledge gives the post height.

Setting the platform

With both rim joists installed, fill in between with joists previously cut to length (their length is the width of the landing less 3 in. for the two rim joists). Two 16d common nails through the rim joist and one up from below through the ledge secure the infill joists.

The final step in building the landing is adding the subfloor. I almost always use ¾-in. subflooring—either oriented strand board (OSB) or plywood. To eliminate the chance of squeaks, secure the subfloor with adhesive and 2-in. screws spaced at 6 in. along the perimeter and at 12 in. in the field.

Laying Out the Stringers

Just as shown in the demonstration photos on p. 23, I lay out the stringer using a framing square and stair gauges. Because this stair has subtreads, it's important to remember to deduct both their thickness and that of the finished tread from the bottom riser cut. Specifically, the subtreads measure ¾ in. thick, and the finish treads will be 1⅛ in. thick. Accordingly, I deduct 1⅞ in. from the bottom riser (but not on the upper flight because the landing doesn't have the finish floor installed).

The landing's finish floor will be ¾ in. higher than its subfloor. Since the stringers for the upper flight will rest in part (more on this soon) on the subfloor, the bottom riser must be ¾ in. higher to account for the finished

A square and pencil mark rise and run. **LVL stringer material has a waxy surface to protect it from moisture. That also makes it a little harder to pencil lines on the material, so bear down with your writing implement.**

floor. So, adjustment for this riser cut is 1⅞ in. (the sum of the subtread and the finish tread), minus ¾ in. for the finish tread, or a deduction from the unit rise of 1⅛ in.

There's an important caveat when using subrisers and a hangerboard. The hangerboard takes the place of what would be the topmost subriser. All of the other subrisers will rest

on the tread below, effectively reducing that tread's run. No problem though, as the subriser at the front of that tread effectively increases the tread's run by a like amount, putting you back where you want to be. However, the hangerboard goes behind the top plumb cut on the stringers, which places it behind where it would go if it were a true subriser. That means the topmost tread cut ends up being the thickness of a subriser too long. It's an easy fix—all you have to do is remember to move the plumb cut at the top of such stringers forward by the thickness of one subriser. In other words, if all of the other treads are cut out to a unit run of 10 in., the top one should be cut to 9¼ in., assuming ¾-in.-thick subrisers.

To support the bottom of the upper flight's stringers, I could have extended the landing farther back so that a level cut across the stringer was fully supported. That's time-consuming, wasteful of materials, and not necessary. Instead, I extended the landing back only 2 in. Likewise, the level cut on the bot-

Hangerboard Detail

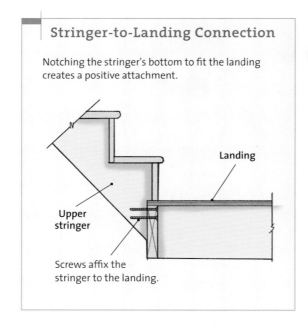

A hangerboard is a piece of ³/₄-in. plywood used to attach the top of the stringers to the upper-floor framing.

Screws affix the hangerboard to the stringer.

Stringer-to-Landing Connection

Notching the stringer's bottom to fit the landing creates a positive attachment.

Landing

Upper stringer

Screws affix the stringer to the landing.

Notch the stringer to fit the landing. Instead of a full level cut at the bottom riser, the stringers for the upper flight are notched. The plumb cut butts to the side of the landing's end joist.

Split the line with the sawblade. For rough stringers, eyeballing the cut using a circular saw is accurate enough. Don't cut past the intersecting riser or tread line, though, as that weakens the stringer.

Finish the cuts with a jigsaw. Although a handsaw would work, a jigsaw with an orbital setting is worth bringing from your truck. LVL material is harder to cut than lumber stringers, giving the handsaw user more of a workout.

tom of the upper flight extends back only 2 in., where it meets a plumb cut extending to the bottom of the stringer. You'll see how this is attached to the landing coming up.

Cutting and fitting the stringer

The cuts in the stringer for the treads and risers are made using a circular saw and a jigsaw. I use both because the circular saw can't complete the notch without overcutting on one face of the stringer. That unnecessarily weakens the stringer. Instead, cut to the rear intersection of the tread and riser, leaving the cutout hanging by a thread. When you've completed all of the cuts with the circular saw, go back and finish them with a jigsaw. A handsaw would also do the job, but the LVL stock I use for stringers is tougher than lumber straight from the tree, so I prefer the jigsaw.

After the first stringer is cut out, carry it to the stairwell for a test fit. Once it's in place, check the tread cuts for level, and verify the fit at the top. This involves measuring from the upper floor's subfloor to the top tread cut. This distance should equal the unit rise plus the combined thickness of the tread and subtread, less the thickness of the upper finish floor. As

long as you've thought this through, everything should check out fine.

Assuming a good fit, scribe a pencil line on the studs below the stringer. This will serve as a guide for installing a 2x4 spacer on the wall.

Test-Fitting the Stringer: What Can Go Wrong?

If the treads pitch forward, odds are that you forgot to shorten the top plumb cut by the thickness of one subriser. If you don't put two and two together and shorten the top of the stringer, you'll probably make some adjustment to the bottom of the stringer until things level out. You then won't figure out what happened until you attempt to place the topmost subtread and find that it's too short. The solution even at this point is simple: Just cut and add another subriser over the face of the hangerboard.

If the treads pitch backward, you may have forgotten to add the shim that represents the hangerboard.

If neither of those solutions fit, go back to square one and check your math and your planning. Check the unit rise and run on the stringer and verify the landing's location. Sometime during this process, the problem should become clear.

Mark the stringer on the wall. While test-fitting the first stringer, the author scribes its bottom to the wall studs as a guide. When test-fitting this stringer, it's important to add a shim the thickness of the hanger-board to the top plumb cut.

Use the first stringer as a template. There's no point to laying out each stringer using a square when you've got one that fits right. Clamp it in place and trace its shape on the remaining stringer stock.

Make a spacer. Mark a 2x4 to fit along the bottom of the first stringer. It will be used to space the stringer away from the wall to create space for drywall and a finish skirtboard.

The spacer leaves room for drywall and skirtboard. Aligning the spacer's bottom with the line previously scribed on the studs, nail it to the wall. Two 16d common nails per stud are sufficient, although screws would also work.

The spacer creates room for the drywall to slide behind the stringer, plus space for a finished skirtboard (see chapter 3) and some shims.

With the fit checked and the line scribed, take the stringer down and return it to your sawhorses. Lay it atop the next piece of stringer stock, align its uncut edge with an edge of the underlying stringer, clamp them together, and trace its shape onto the next stringer. Repeat

for as many stringers as you're using, which is rarely more than three, particularly if you're using LVLs.

While you've got the stringer on the horses, lay a long 2x4 along its bottom and trace the top and bottom cuts from the stringer to the 2x4. That's the spacer, which you then carry over to the wall and nail with two 16d commons per stud.

Installing the Stringers

A hangerboard affixes with nails or screws into the top plumb cut of the stringers. It extends the width of the stringers and is tall enough to reach to the bottom of the plumb cut from the top of the upper floor's subfloor (see the top drawing on p. 30). When the hangerboard is placed, nails or screws hold it to the floor framing at the head of the stairwell. It's a simple, effective method of attachment. In most cases, it hangs down below the floor framing far enough that the stringers can be placed individually and fastened from behind.

Dealing with deep floor joists

It's not uncommon these days, however, to find floors framed with engineered lumber that's deeper than conventional sawn lumber. The floor shown here is 16 in. deep, which leaves no access for securing stringers through the back of the hangerboard. Consequently, they all had to be affixed to the hangerboard before it was placed. Not wanting to do that on the floor and be stuck manhandling several hundred pounds of stringers into place, I had to come up with a plan.

I tacked a piece of 2x6 to the upper floor a couple of feet into the stairwell and leaned the stringers against it. The notches at the bottom of the stringers hooked onto the landing and kept them from skidding. (You might think this arrangement is a little esoteric, but I've used this trick at other times to prop up a set of shop-built stairs during installation. It's not at all uncommon to have to maneuver stairs or their components in increments. Having a means to secure them temporarily allows you to move around. The key is to plan out the moves before moving anything.)

What's holding up the stringers? **A scrap of 2x6 temporarily nailed to the upper floor holds the stringers away from the head of the stairwell, providing space to nail the hangerboard to the stringers from above. A cleat across the face of the hangerboard aids in keeping them properly positioned.**

No access after installation. **The view of the installed stringers shows the need for fastening the hangerboard to the stringers before dropping the assembly into place, when very deep joists are used.**

Remember that spacer? **The outside stringer is nailed to the 2x4 spacer, further strengthening the assembly.**

Strong stringers. **Even before the hangerboard was nailed into place, the stringers easily supported the author because they were captured at the top by the floor framing and at the bottom by the notches that go onto the landing.**

Block between the stringers. **Cleats nailed to the back of the landing between the stringers provide nailing. One is in place, and the second cleat comes after the stringer is fully nailed to the first.**

With the stringers propped up, I was able to clamber up to the second floor and fasten the hangerboard to the stringers. Because I knew I wouldn't be able to see exactly where the stringers were on the hangerboard and because that location determined the proper installation of the stringers, I screwed a cleat to the hangerboard at the line where the tops of the stringers had to go (see the middle photo at right). I knew I'd be able to feel them butt against that. Side alignment was simple, as the stringers had to be kept flush with the edge of the hangerboard. That, I could see. That left the center stringer, and a vertical alignment cleat fixed that.

I used 16d commons again, although screws would also work. If you're questioning my use of nails here, just consider that in Douglas fir lumber, a properly driven 16d common is good for nearly 100 lb. of shear load. I used four in each stringer. Plus, one stringer was nailed to a wall and another supported by a wall. The notches at the bottoms of the stringers hooked to the landing. This stair isn't moving.

With all the stringers attached to the hangerboard, all I had to do was pull the nails holding the 2x6 to the second-floor framing and slow

Overkill isn't enough. Although the hangerboard is more than adequate to support the stringers, cleats are a belt-and-suspenders approach that will also serve to support the upper subtread.

Engineered lumber isn't perfect. Blocking between the stringers keeps the spacing between them the same as at their tops and bottoms.

Hangerboard, the easy way. The bottom flight is wider than the upper, so additional stringers were used.

the fall of the stringers and hangerboard. The hangerboard was nailed to the head of the stairwell with, you guessed it, 16d commons.

The notches in the bottom of the stringers do a great job locating the stringers, but I like to use a belt-and-suspenders approach, particularly with the center stringer that enjoys no help from walls. A 2x6 cleat nailed to the landing between the center and the outside stringer provides a target for nailing through the center stringer. A cleat to the other side for good measure helps to keep the stringer from twisting under load. Similar cleats were installed at the top, with the added benefit of providing support for the topmost subtread.

Even though LVL material is straighter than sawn lumber, it's not perfect. These stringers were about 13 ft. long, and no right-thinking carpenter expects any wood product to be perfect at that length. Some 2x4 cleats nailed to the stringers lined them up nicely.

The bottom set of stringers also depend on a hangerboard. Since they had only three risers, they were a little easier to place.

Installing Treads and Risers

Much of what gives stairs their strength is that the components work together as a unit. One stringer alone, if your balance is good, will get you from floor to floor. Two would be better, but they'd still flex as your weight shifted from foot to foot. Add treads, and you have the most basic of stairs. They'll flex much less underfoot, as the treads help to distribute point loads between the stringers. Depending on their stoutness, you may still feel the treads themselves flex, though (and I guarantee that you'd feel ¾-in. plywood subtreads flex).

Add risers (or subrisers) into the mix, and they act like beams to spread the load between the stringers. Affix them so as to support the front and back of the treads, and the additional stiffness is nearly magical. It's amazing the stiffness that ¾-in. plywood subrisers bring to the table. The point here is that I would never leave ¾-in. subtreads in place without the support of subrisers. To do so would be dangerous,

The techniques shown here are exactly as I'd use to make a set of utility stairs. Utility stairs would have no subtreads or subrisers. What I use here as subrisers would be the end product on a utility stair. In place of subtreads, I'd use the real McCoy, most likely 2x12 stock with its leading edge rounded over and overhanging the lower risers per code. So far as calculations go, the only difference is that when figuring the height of the bottom riser, there's no subtread thickness in the equation, only the thickness of the actual tread.

Squeak-Free Stairs

The most common complaint people have about stairs is squeaks. Squeaks happen when two pieces of wood rub together. It's hard to overemphasize the importance of gluing and screwing everything that happens from this point on. I use PL 400® construction adhesive, not because I think it's the best brand but because it's readily available and it does the job. Other brands probably work well, too. For screws, I use 2-in. coated deck screws. They don't snap off like drywall screws, and I can get them with square-drive heads that rarely allow the driver to cam out. Square-drive heads are as superior to Phillips heads as LVL is to sawn lumber. Even better, they don't cost more. Try them out.

as somebody carrying something heavy up the stairs would doubtlessly break through. Neither you nor your insurance carrier would want that. If you want to skip the subrisers, fine. But use 2x-something treads if you do.

Making treads and risers

The first move here is to figure out how many treads and risers you need and to rip sufficient amounts of stock to size. Now, for most framing uses, I hold OSB and plywood to be equal. Not for subtreads and subrisers, though. OSB is probably plenty strong enough, but it won't take a screw from the side without splitting. Plywood will. I'll occasionally use up OSB scrap for subrisers, if that saves ripping into a fresh sheet of plywood, but that's all the use I have for it on stairs. I rip all of the subtread stock parallel to the face grain of the plywood—that's the strongest axis. If it allows less wasteful use of the material (and it often does), I'll cut the subrisers perpendicular to the grain. Often, I end up with some subrisers cut either way.

Because these subtreads and subrisers will serve as substrates for the finish treads and risers, being mindful to rip straight edges (a tablesaw is best) and carefully installing these rough materials will ease the finish part of the job. If anything, make the subrisers and subtreads a little narrow. The subrisers can be pulled up to be flush with the tread cuts, but if they're cut too wide, they sit proud and prevent the subtread from properly seating on the stringer. And if the treads are too wide, they'll overhang the subriser and interfere with installing the finish risers. To speed things along, cut all the subtreads and subrisers to length before beginning installation. Use a miter saw to ensure good, square cuts.

The final touch: subtreads and subrisers

1 Install two risers first. Bed them in construction adhesive and be sure they're flush with the tread cut above.

2 Construction adhesive prevents squeaks. Everywhere that a subtread contacts other parts of the stair needs to be liberally coated with it.

3 Screw the subtread to the subriser below about every 6 in. To avoid splitting the subriser, predrill for each screw. Use at least two screws through each subtread into each stringer.

4 Screw each subriser to the tread in front of it. Predrilling is important here, as a split tread has no strength.

Set the saw to bevel the tops of the studs. **Those stringer cutouts are good for more than firewood. They've got the angles needed to set up saws to fit cuts to the stair.**

Start the wall below the stair. Nail a 2x4 to the bottom of the stringer to provide full nailing for the studs to come. It's possible to skip this 2x4, but five minutes spent on this step eases the rest of the wall construction.

Hold the studs plumb, and mark them in place. **A couple of toenails secure each stud. Work from the bottom of the stair toward the top to leave yourself working room. And don't drive the tops of the studs forward while setting the nails or you'll wedge up the stairs.**

Glue and screw treads and risers to stringers

Working from the bottom up, start by installing two risers. Squeeze a generous quantity of construction adhesive onto the riser cuts of the stringers, and run in two screws per stringer. Keep the top screw at least 1 in. down so as not to split the stringer stock.

Bed the treads generously in construction adhesive. A bead run along the bottom face of the upper riser, the top edge of the lower riser, and on each stringer ensures a squeak-free assembly. Don't be shy with the screws, either. Take the time to drill pilot holes for screws going into the edge of the plywood. Although this will seem inefficient, I climb down from the stair and walk behind it to drill and screw through the back of the subriser and into the subtread before it. It's important to do this before the adhesive starts to set up. If the weather's cool, you've got more time, so you might be able to run around and screw two or three at a time. Or, if your back is more limber than mine, you can do this operation by reaching over the top.

Building the Support Wall

Because of all the running around I do, I wait until after all of the subtreads and subrisers are installed to build the wall under the inside of the stairs. This is straightforward framing. Snap a line between the plumb point on the floor (remember that point from the start of this chapter?) and the post supporting the landing. Cut and nail down a 2x4 plate and lay it out on 16-in. stud centers. Then, setting the bevel on your circular saw using one of the cutouts from the stringer, affix another 2x4 to the stringer as a top plate, mark the studs in place, and build a wall.

40 | Fitting the Skirtboards

44 | Notching the Inner Skirtboard

46 | Fitting the Risers and Treads

49 | Attaching the Treads and Risers

52 | Making Returned Treads

Finishing Rough Stairs

The last time you saw these stairs, they were just rough-framed. Now it's time to add the finish skirtboards, risers, and treads. This is pretty fussy finish work. It's likely that the rough stairs aren't perfect, even if you built them yourself. All bets are off if they were built by the framing contractor. This is the interface between rough and finish work. Not only do you need to fit treads and risers nearly perfectly between and around skirtboards, but also the riser cut on the inner skirtboard is a compound angle that forms a miter joint with the riser.

And, you're doing all of this with good and expensive materials. In the example shown here, I made the skirtboard from primed, finger-jointed 1x12 pine, which saves the painter a step. It's also nice material to work with, relatively stable, straight, and flat. Another common choice for work that's to be painted is poplar. Where I live, the cost of the two is comparable, so I give the painter a break.

The treads are oak, with returns on one end. I can buy returned treads, but they're pricey enough that I find it profitable to make my own (see "Making Returned Treads," pp. 52–57). If the stairs are to be carpeted, and the owner requests it, I can make the treads of a less expensive material like poplar. I've even seen medium-density

Start with Cleanup

You've come back to the house where you built the rough stairs months ago. The hardwood floors are probably down, and the drywall is up. Odds are, there is drywall dust in all the nooks and crannies of the stair and maybe even some scraps jammed in here and there. So, job one is cleanup. So much for the romantic job of master stairbuilder. I go so far as to vacuum the stair. Drywall dust is insidious stuff—it gets in your tools, your hair, the back of your throat where it wreaks who knows what sort of medical havoc. It interferes with glue joints, and big chunks prevent you from properly seating finish stair parts. Get rid of it.

Finish treads and risers transform a rough stair. **The plywood subtreads and subrisers got the house through construction; now it's time to dress the stair with finish-grade material.**

fiberboard (MDF) used for treads and risers. Not that there's anything wrong with that; I just don't like how dusty and heavy MDF is to work with.

Fitting the Skirtboards

Get started by figuring out how high the wall-side skirtboard will go. It's important to do this so that you can draw out the end cuts of the skirtboard in place. Lay a level along the steps and in contact with the wall, and trace a pencil line along its top. That's a benchmark you can easily re-create anywhere along the stairs. You have to make sure to keep the bottom of the skirtboard no higher than the intersection of the risers with the backs of the treads. Measure up from that intersection on one step, and mark the wall at the width of the skirtboard. Measuring the distance between this point and the benchmark line gives you a dimension you can mark up from the benchmark farther down. Connecting these dots with a pencil line

gives the angle of the stairs and the location of the skirtboard's top. Make a similar set of lines at the top of the stair.

From these lines, you can use a protractor and a level to find the angle of the skirtboard's plumb and level cuts. Moving the plumb cut at the bottom of the skirtboard forward makes it shorter. I usually make it the same height as the base molding that will be used on the landing. Penciling the top and bottom plumb cuts on the wall also provides two points to measure from to determine the overall length of the skirtboard. I usually cut the skirtboard with a circular saw, but a big miter saw will do it, too.

Attaching the skirtboard

Once you're happy with the skirtboard's fit top and bottom, it's time to nail it home. It's tempting to simply nail the skirtboard directly over the drywall and call it good, but this may complicate installing the treads and risers. A

Marking the skirtboard on the wall

1 Lay a level across the tops of the sub-treads to guide a pencil line that serves as a benchmark for the skirtboard.

2 Mark the skirtboard's approximate width on the wall, measured from a point that's sure to be covered by the finish treads and risers.

3 Mark the difference between the approximate height of the skirtboard and the benchmark line near the bottom of the stair.

4 Connecting the dots accurately locates the top of the skirtboard, providing a line that can be used in plotting out its length and angles. The top of the skirtboard is marked out in a similar manner.

Plumb the skirtboard.
Shimming the skirtboard
plumb will make it eas-
ier to install the treads
and risers. Shims back up
the skirtboard as well,
taking out any flex that
would allow the joints to
open up.

Nail it off. Secure the skirtboard with three 2¹/₂-in.
finish nails into every stud, shot through the shims.
Use a utility knife to score the shims along the skirtboard, and break them off flush.

Before the skirt, install the newel. Because the inside
skirtboard gets shimmed out, screwing the landing
newel directly to the framing and fitting the skirtboard to it instead makes for a stouter connection.

little time spent now shimming the skirtboard
plumb will pay dividends later. In the hope of
keeping the skirtboard from moving later on,
I shim and nail it to every stud using 2½-in.
trim nails.

Fitting the inside skirtboard

The inside skirtboard has to work with a landing newel. Rather than install the newel on top
of the skirtboard that will have been shimmed
to be plumb, it's better to install the newel

When fitting the newel on a stair like
this, cut it just short enough so that the
finish tread below slides under with a firm
push. This step saves the time-consuming
work of notching a tread.

directly on top of the framing. This makes for a
stronger attachment, and fitting the skirtboard
to the newel isn't difficult. (You might have to
skip ahead now as the mechanics of fitting a
newel are covered in chapters 8 and 9.)

Laying a straightedge such as a level along
the stair provides a base for measuring the
angle at the top of the skirtboard. Marking the
ceiling where the line of the stair intersects
it provides a starting point for measuring the
skirtboard. I don't try to measure it perfectly—
that rarely works as well as hoped. Instead, I'm
content with an approximation that I know
is a little long. I screw a couple of cleats to the
wall to hold the skirtboard in place temporarily. The skirtboard should be held about ¾ in.
above the front of the treads, or riser miters
won't meet the tread cuts at a point. I then
mark the lower end for cutting, leaving a little
extra stock for fine-tuning the fit.

Fitting the inner skirtboard

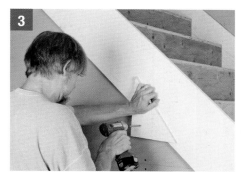

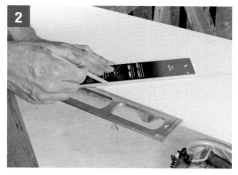

1 Using a level as a platform, measure the angle of the top cut where the skirtboard intersects the ceiling. The trim board makes a better joint with the skirt than would the drywall, and also offers more support for balusters later on.

2 A 12-in. protractor is an ideal tool for transferring the upper angle to the skirtboard stock. Flipped end for end, it easily does the entire board's width.

3 Screw a temporary cleat to the wall to hold the skirtboard for fitting. The skirtboard actually rests on a block of scrap hidden behind the big triangle, which pivots out of the way to allow the skirtboard to be removed and replaced as needed.

4 Start by cutting the skirtboard too long. Lap the newel, then scribe the skirtboard to the newel from behind. It may take several fittings to get a good mate (proving the value of the pivoting cleat).

5 After the initial cut, eyeball the gap at its largest and hold the pencil point at about the same distance from the newel to scribe for a perfect fit.

6 Plane with the grain. A sharp block plane quickly takes the end of the skirtboard down to the scribe line and leaves a nice, straight mating surface.

Notching the Inner Skirtboard

With the skirtboard fit to the newel and the
ceiling, now comes the part that's most likely
to scare the uninitiated—laying out and notching the inner skirtboard with its compound-
angle miter cut. Good news: This is easy. Using
a flat piece of scrap that's the same thickness
as the riser (scrap skirtboard works well) and
at least a couple of inches taller than the riser
height, I make a version of what's known to
siding carpenters as a "preacher". A preacher
is essentially a board with a notch cut from it
so that it fits around the material you're installing. With the stock held in place, the preacher
straddles it, bearing against the surface you're
fitting to and providing a guide for striking
a line.

Making a Preacher

A "preacher" is a site-built marking jig. I make a preacher from
scrap 1x12 using a miter saw to create good, straight, and square
cuts for the sides of the notch, which should be a little wider
than the stock it's to straddle. After cutting the sides straight
and square with the most precise tool in my arsenal, I chop the
scrap from the notch with a chisel and hammer.

A short level keeps the preacher plumb. **Because the
preacher is the same thickness as the riser stock, its
face represents the point of the miter on the skirtboard. To make accurate marks, the preacher must be
pushed tight to the subriser.**

Marking for risers and treads

With the preacher's notch straddling the skirtboard and its back held to the subriser, hold
the preacher plumb. Then mark the face of
the skirtboard using the face of the preacher
as a guide. Because the preacher is the same
thickness as the riser stock, this pencil mark is
exactly where the outside edge of the miter on
the skirtboard should go.

Because the drywall runs over the rough
stringer, and the skirtboard runs atop the
drywall, there's a drywall-wide gap between
the rough stair and the inner skirtboard. The
preacher bridges that handily when marking
the risers, but marking the tread cuts is a little
tougher. You could make a larger preacher
for that purpose, but that would be an awkward tool. And, the longer the notch in the
preacher, the more likely it is that the material
to the sides of the notch won't stay in place.
That wouldn't make for accurate layout.

Instead, I just lay the preacher flat on the
subtread and mark its top on the skirtboard
as a reference line. When I take the skirtboard
down for cutting, I align the preacher on these
reference lines and mark along its bottom for

Mark a reference line for the treads **along the top of the preacher. There's a gap between the stringer and the skirtboard because the drywall runs between them below. This gap prevents directly tracing the subtread to the skirtboard.**

Mark the cut lines for the treads. **After taking down the skirtboard, place the preacher along the tread reference line and mark its other side, which is level with the subtread, for the tread cut line. Scribble out the initial mark to avoid mistakenly cutting to it.**

The riser cuts are compound angles. **A sliding compound-miter saw set to the riser angle and a 45° bevel handles the cuts, which are made from the face of the skirtboard.**

Cut the treads from the stair side of the skirtboard. **Ease the blade in, starting from the back of the tread, and watch so it doesn't overcut. Note that the reference lines are scribbled out to avoid confusion.**

the tread cut line. The tread cut is not as critical as the riser cut. The riser miter will show, whereas the tread cut will be hidden by cove molding.

Cutting the skirtboard

My favorite tool for cutting skirtboards is a sliding compound-miter saw, but it's got to be able to bevel 45° both right and left to make the compound angle cuts needed for the riser cuts. Clamp the stock to the table for the miter cuts. Any movement of the stock up off the saw's table will alter the cut.

To avoid having to repeat the setup, I set the saw to a 45° bevel and make all the riser cuts first. The angle of these cuts should be consistent, so it's a matter of matching the saw's miter angle to the riser layouts, locking it in place, and making the cuts. By the way, cuts with a sliding miter saw should never be made by pulling the saw toward you. That's climb cutting, where the blade contacting the stock is spinning in the opposite direction that the saw is moving. In that circumstance, the blade can grab the stock and leap backward, ruining only the cut if you're lucky. Always pull the saw head out, align the cut, lower the blade into it, and push.

After all the risers are cut, set the bevel back to 90°, adjust the cut angle to that of the treads, and finish up by cutting the notches. Before I invested in a large compound miter saw, I mitered a lot of skirtboards using a circular saw and a shooting board (see pp. 54–57). If you go that route, you'll need both left- and right-blade circular saws. Right-blade saws will miter right-hand skirtboards but not left-hand ones. The opposite is true of left-blade saws. It's a bit of a mind bender, and at the time I bought my second circular saw, I had a job convincing my wife that it was need and not my burgeoning tool addiction that led me to buy a second saw. The need had been made clear to me, however, as I'd just had to miter an oak skirtboard using a handsaw.

Don't worry about completing the cuts on the skirtboard with the miter saw. I finish with a handsaw and use a chisel to clean up.

Cedar shims and a level keep the skirtboard plumb. It's not uncommon for the notches to curl a bit after they're cut. Straighten them out and nail each corner fast through the shims into the stringer.

Fitting the Risers and Treads

Because the skirtboard is the transition to relatively rough work—the underlying framing and drywall—I never simply slap it up in the naive assumption that everything below is plumb and level. That's even the case when I built the wall below, as it's not immune to the ravages of lumber shrinkage and I'm not immune to doing imperfect work. Throw a drywall crew into the mix, and pretty much anything is possible.

Place the skirtboard back on its temporary supports, double-check its fit top and bottom, and nail its bottom in place. Then nail the top of the skirtboard in place, first checking along each riser miter for plumb. Shim behind here as needed for support using cedar shingle shims. The closer to plumb you can keep these cuts, the easier it will be to fit the risers.

With a sharp paring chisel, cleaning up the cuts on the skirtboard takes only a few minutes. Nailing up the skirtboard beforehand provides a solid foundation for the chisel work.

You can make your own version of the Tread Template with two overlapping pieces of plywood. Hold the plywood pieces in place much like the store-bought tool, and clamp or screw them together. It's more awkward to use, but it is cheap.

Laying out the treads and risers

If you've managed to keep all the subtreads, subrisers, and skirtboards plumb and level, all you'll need to do is measure, mark, and cut the risers and treads square. Let me know if you pull that off. I find that there's usually something imperfect. Now, there are levels of imperfect. What I'm talking about here are imperfections on the order of being half a degree, or even a degree, out of plumb and level. It's not important on a practical level, but to a trim carpenter, a miter that's open by a degree is like staring into the maw of a Florida gator. It's just not something you want to see most days.

The best way I've found to avoid the gator here is to use a specialty tool, the Collins Tool Company's® Tread Template. It consists of a pair of 12-in.-long aluminum straightedges with integral clamps. These fit a piece of 1x4 (look for one that's straight and flat) that's cut an inch or two shorter than the distance between the skirtboards. Mark one end of the 1x4, and always keep it on the right or the left. This avoids accidentally flipping the template over and cutting the stock backwards.

To use the Tread Template, make sure the ends of the straightedges sit on the subtread or the subtread and the tread cut on the skirtboard. Adjust and clamp the side against the

A special tool aids in fitting. The Tread Template's length is adjustable, and its ends are set to fit the skirtboards on either side. Mark one side of the 1x4 that joins the metal ends to avoid inadvertently marking the stock backwards.

Perfection is pretty rare. It might seem that fitting a riser to a newel should take a straight 90° cut. Often though, something is out just a tiny bit. Finding length and cut angles with a template guarantees a good fit.

outer skirtboard first, so that it can be seated firmly against that skirtboard as you're adjusting the straightedge along the inner skirtboard's riser miter. The Tread Template also shines at establishing the shape of risers or treads that fit entirely between closed skirtboards or to a newel post.

With the template set, carry it over to the riser stock, and mark the stock for cutting. One end receives a straight cut to fit to the outer skirtboard, and the other end is beveled at 45° to fit the mitered skirtboard. That's the bevel I'm talking about, the angle at which the blade

Be sure to start in the right spot. **When marking stock to be cut, align both points of the template that rested on the reference surface—the subtread in this case—on the appropriate edge of the stock.**

Align the open end first. **For treads, start the template alignment at the outside of the skirtboard, paying particular attention to the back where a tread would return.**

intersects the saw table. The miter angles—how the blade intersects the fence—should be darned close to 90°. It should be off by no more than a degree or two. The quickest way to set this angle is simply to eyeball the blade along the cut line.

Measuring the treads

The treads are measured in much the same way, although I don't worry at this point whether the returned end will be perfectly square to the inner skirtboard. Because the treads are made and the returns applied in the shop (see pp. 52–57), I know they're square. If the risers and skirtboard differ a little, there's not much to be done, and the only place the discrepancy would be noticeable is if you're looking very carefully at the bottom of the tread. One of the benefits of tread overhangs and cove molding is that they hide such minor transgressions.

I do, however, pay attention to aligning the template on the outside skirtboard. To do this, I start at the other end, lining up the point of the template on the point of the miter at the back of that tread on the inner skirtboard. This puts the template the thickness of a riser from the subtread, and I use a scrap of riser stock

Eye the alignment along the tread returns. **The starting point for the template when marking a tread is where the return passes the back of the tread. The template should line up along the tread return; if it doesn't, that's a signal to double-check yourself.**

to similarly position the template at the outer skirtboard. I mark the tread, placing the point of the template that aligned with the skirtboard miter on the point where the tread return passes the end of the tread, and make the cut on a miter saw. A circular saw will also work.

Attaching the Treads and Risers

Unlike many other repetitive stairbuilding operations, I don't cut all the treads and risers to size before assembling them. That's because the act of assembly can change the dimension of the next set of tread and riser, making the precut parts fit badly. So, I measure and cut the lowest riser and nail it into place. In the case of this stair, the first step was a bullnose step (as shown in chapter 6). This is a pretty common occurrence, and bullnose steps come with their bottom riser in place. That's the only practical difference as far as installation goes.

Installing treads and risers

Since the subrisers leave no access for screwing through the back of a riser into the tread before it, each tread and upper riser go in as a unit. Screw and glue the two together before placing them, using yellow glue and 1½-in. screws. This is a good time for a reality check as well. When you hold the riser up to the back of the tread, the square ends should be flush, and the mitered end of the riser should just kiss the inside of the tread return.

Spread out several beads of construction adhesive on the subtread, run a little yellow glue on the riser miter, and set the tread in place. The tread and riser should butt tightly to the outer skirtboard. Pinch the miter together, and shoot in five or six 1½-in. finish nails. A few more nails secure the riser to the subriser.

Most of what keeps the tread in place for the long run is the construction adhesive, but the treads do need to be held down until the adhesive sets. My favorite approach is to stand on the tread to hold it in place and have an assistant run in six or eight 1½-in. screws from below through clearance holes drilled

through the subtreads. That way, there are no holes in the top of the tread to fill. That's not always possible, though. Sometimes, there's no access. Other times, I'm working alone. In these cases, I stand on the tread to bed it and shoot in some 2-in. finish nails. I use T-head finish nails, which are a little less noticeable if I shoot them in so the head is parallel with the grain. When working alone, I use just enough nails to keep the tread from moving, then run around and screw it from below. Drilling oversize screw holes through the subtread is particularly important here. If you don't, the screw is likely to engage in the subtread and push the tread out of place before engaging, thus pulling the tread down to what's probably the wrong place.

Fitting treads and risers to a newel

Treads whose risers fit to a newel are installed in much the same way. If you were smart and cut the newel to allow it, the tread just slides right in. A few 2-in. finish nails secure the riser to the subriser.

Forethought pays off. The newel was cut short enough for the tread to slide right under. That's a lot quicker than notching the tread to fit the newel.

Installing the treads and risers

1 As with any tread and riser, the upper riser is screwed and glued to the back of the bullnose tread prior to installation. The bottom of the riser should be flush with the bottom of the tread, and the riser's ends should align with the tread return and the wall end.

2 Use a generous amount of construction adhesive on the subtread before placing the tread. It should seat tightly against the outer stringer, and the riser miter should align.

3 Secure the miter joint between the riser and the skirtboard with half a dozen trim nails. Shoot several more through the riser into the subriser behind.

4 The adhesive will do the lion's share of holding the tread down, but some human weight and a few finish nails keep it in place until the glue sets. Where there's access, screws from below leave no holes to be filled.

One trick to preventing a nail being hammered through hardwood from bending is to pinch it between your thumb and two or three fingers. The amount of rigidity this adds is surprising, and it's borne out by how rarely nails driven in this way bend.

Lay out landing tread to fit around a newel by holding it in place and marking the notch. After cutting and test-fitting the landing tread, run a couple of beads of construction adhesive below and set it in place. Landing treads get stepped on a lot, and there's not much material to fasten. I don't use gun nails on the landing tread, preferring to hand-drive traditional 8d finish nails. If I'm nailing near the end of the board, I'll drill a pilot hole for the nail. I don't bother in the middle, though. I drag the nail over a lump of beeswax or paraffin, and blunt the tip with a hammer blow to minimize the risk of splitting.

Adding the cove

The last step is adding the cove below the treads. This is straight-up trim carpentry. Hold each piece that runs below the treads in place, and mark it to length for mitering. Nail that up, then hold the cove that runs below the tread return in place, and mark its length. The only trick I have to offer here concerns how to end this short piece of cove. A purist would miter and return the end, and I've done that. Cutting a return from $1\frac{1}{16}$-in. cove molding, and then finding the return from wherever the miter saw flung it, is time-consuming. This is a niggling detail that's not worth the result.

I'm not the first carpenter to figure that out, and many of my colleagues simply chamfer the end of the cove at 45°. That looks cheesy, in my less-than-humble view. Instead, I chamfer the end of the cove at 35°, just up to the flat that faces outward. Distinguishing this chamfer from an actual return takes a close

Mark the landing tread to fit the newel. Hold the landing tread just behind its installed location to mark the notch that will fit it around the newel.

Fasten the landing tread for the long term. Use plenty of subfloor adhesive under the landing tread, and nail it down with hand-driven 8d finish nails. Drill pilot holes for the nails near the ends.

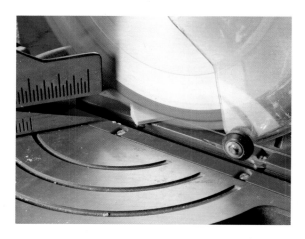

Cheating isn't always bad. Chamfering the cove for below-the-tread returns at 35° yields a look that closely mimics a full return, at about a quarter of the labor.

look. Only one of my customers has ever commented on it, and that was a builder who had been a regular customer for years. He asked how many of his stairs I'd trimmed that way. When I told him all of them, he said, "Carry on."

Making Returned Treads

An awful lot of trim carpentry is designed to hide end grain. Miters frequently exist to hide end grain, as do molding returns. There are sometimes secondary reasons to hide end grain, but the main reason is that end grain finishes differently than face grain.

On notched-stringer stair treads, hiding end grain means mitering the forward 1 in. to 1½ in. at the end of a tread and adding a return that runs the length of the tread. This return usually runs past the back of the tread by the same distance as the nosing overhang. It's a deceptively simple-looking thing. First off is the issue of attaching the return to the tread. The tread's end grain won't glue well to the return, and even if it did, seasonal movement would soon rip the joint apart. But even before you worry about seasonal movement, you've got to miter the tread and crosscut up to that miter, leaving a clean enough cut for a good-looking joint.

Making the miter cut

The trouble with the miter cut is that it has to be dead accurate, but there's no way to make the cut with a miter saw. It has to be cut from the end of the tread because its bottom has to be flat. This was a vexing problem when I first encountered it. Then I remembered an article in *Fine Homebuilding* magazine where carpenter Larry Haun would clamp all the rafters for a house together on edge and gang-cut the bird's-mouths with a big circular saw. Stacking all the rafters together made a platform the circular saw could ride on. All Larry had to do was set the saw to the right angle and depth and cut a straight line. Likewise, by stacking all of the treads I'm making at one time on end, the treads themselves could serve as the platform. All I would have to do is set the saw to 45° and the depth of cut to the width of a tread return and cut a straight line. And that's essentially how I do it.

You have to cut the treads to the same length exactly so that when they're standing next to each other their ends are flush. Clamp them together so the edges to be mitered are flush, and check the stack for square all ways. Often, I'll even them up the last little bit using

Hiding the end grain. A mitered return hides the end grain of a tread, providing a finished look. The tricky part is mitering the front of the tread.

Mitering the tread

1 Make the treads into a solid block. Clamp the treads for mitering, and check their ends and sides for square. If necessary, use a dead-blow hammer to coax individual treads into conformance.

2 The ends almost never end up perfectly aligned, but a few minutes with a low-angle block plane puts it right.

3 With the shooting board placed on the upside-down saw, use a piece of return stock to set the depth of cut. It must equal the width of the nosing overhang, usually 1¼ in.

4 Screw the shooting board to the ends of the treads and make a slow and careful cut.

a block plane. In years past, I would then screw a scrap of plywood to the ends of the treads to act as a fence for a circular saw. Recently, I've started using a shooting board instead.

The shooting board I use for mitering treads is fairly short, about 16 in. long, and I screw it to the squared-off stack of treads so that the sawblade will just split the outer corner of the treads. This ensures that the cut is consistent and of the right depth. But before screwing it to the treads, I place it on the upside-down circular saw to set the depth of cut. I do take care to set the saw's bevel as close to 45° as possible, using a Speed Square to confirm the setting. Circular saws being what they are—framing tools—sometimes the bevel is off a little. This isn't a big problem, as fine adjustments to the miter cut on the return are easily done.

One other advantage to stacking the treads for this miter cut is that splintering where the blade exits the cut is minimal. That point on all but the last tread is supported by the next tread, and a sacrificial bit of scrap clamped to the last tread would take care of that.

Making the square cut

There a couple of ways to make the square cut on the tread. The low-cost way is to use a circular saw and a shooting board, cutting just to where the miter begins. Before making this cut, scribe the tread with a knife and a square to eliminate splintering. I did it this way for years, until a sliding compound-miter saw made its way into my life. That's the tool I use now.

No matter how most of the cut is made, the nature of round sawblades means that there's a small bit of the cut that has to be finished with a handsaw. I'm not so good with a handsaw as to finish the machine cut perfectly, so there's always a little to be cleaned up with a chisel.

Square-cut back to the miters. A circular saw and a shooting board can make a straight cut that tightly fits the return. Alternatively, you can use a sliding miter saw, a radial-arm saw, or a tablesaw set up with a sled.

Splintering is easily avoided. Knife-cut the line of the return before sawing it to make a crisp edge.

Shooting Boards Turn a Circular Saw into a Precision Tool

A shooting board consists of a base and a fence. I usually make the base from ½-in. or ¼-in. plywood or from ¼-in. Masonite®. The fence is made from either ¾-in. or ½-in. plywood. Half-inch plywood is better, as it's thick enough to reliably guide the saw and thin enough to be out of the way of the saw's motor when the blade is down all the way.

Make the fence about 2 in. wide, with one straight edge. The base is about 12 in. wide to offer up plenty of space for clamping. Glue and screw the fence to the base a little farther from the edge than the width of the saw base. The first time the shooting board is used, the saw cuts the shooting board to exactly the distance between the edge of the saw's sole and the blade. Now, the edge of the shooting board is exactly at the edge of the sawblade, preventing splintering on that side. And depending on which side of the stock is the keeper, you can either clamp the shooting board to the stock directly on the cut marks or the thickness of the blade from them (which is how it has to be done when mitering stair treads).

Note that you can't use the same shooting board for square cuts and for bevel cuts. You'll need separate ones for each operation. You can also make a full-length shooting board to take the place of a tablesaw for many operations.

Prop a new shooting board up on scrap for the first cut. This trims it to the width of the saw sole, creating a zero-clearance base that minimizes chips and can be placed exactly on the cut marks or a kerf away, depending on which side of the stock is the keeper.

Different saws and bevel angles call for their own shooting boards. This one is being used to miter a finish stringer.

Splines are quick and strong. **The tread ends and the returns were slotted on a router table for splines, which reinforce and align the joint between them.**

A slot cutter on a router prepares a tread for a spline. **The protruding miter precludes using a fence while slotting the tread. For safety, always use a fence while slotting the return.**

Finish the saw cut with a handsaw. **You can't use a circular saw to complete this cut, but a jigsaw or a well-sharpened handsaw does the job.**

Round over the tread return ends while they're still a block of wood. **Then rip the returns from the block. This avoids tearing out the grain as would happen if the ends were rounded over after ripping.**

Clean up the final handsaw cut with a paring chisel. **Use the machine-cut part of the tread to guide the back of the chisel.**

Making the returns

The return stock itself is easy, just material of the same species and thickness of the treads that's ripped to the width of the tread overhang. I make the returns longer than the tread is deep by the width of the overhang so that they'll pass beyond the upper riser. The only trick to this is in rounding over the end of the return that extends behind the riser. Because this is end grain, and a small piece of stock, doing the roundover after the stock is ripped to width is problematic. It can be done safely when the assembled tread and return are rounded over, but the router bit will always blow out some grain on the end of the return.

To avoid this, I rip the returns out of a wide although fairly short piece of stock. But before ripping the returns, I round over the end grain with the same bit I'll use everywhere else. Then I rip to width on a tablesaw and miter the individual returns on a miter saw.

In addition to tightly fitting to the tread, the return must be flush with the top of the tread and stay that way for the life of the stair. To accomplish this end, I spline-join it, using a slotting bit in a router to groove both the return and the tread. That takes care of alignment but not attachment. For that, I use a Kreg Jig® to drill two pocket holes in the tread. I keep the screws near the ends of the

A pocket-screw jig readies the tread to join to the return. **Place the holes near the front and back of the tread to keep screws away from future baluster locations.**

Pocket screws pull the return tightly to the tread. **The spline keeps the pieces aligned, and a pipe clamp holds the return in place until the screws are set.**

Faster, quieter, and less dusty than a sander, a sharp smoothing plane levels the tread and return with a minimum of fuss. Paying attention to grain runout and planing obliquely prevent tearout.

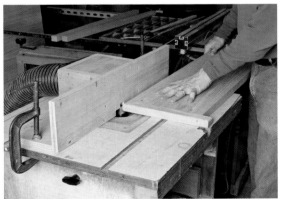

Round the nosings on a router table. **Finish up the treads with the same roundover bit that was used on the return stock. A couple of minutes of sanding will remove the machine marks.**

tread so they're unlikely to interfere when I drill the holes for balusters. When assembling the returns to the tread, I use a little glue at the mitered end but nowhere else. That's an attempt to force seasonal movement to be noticeable at the far end, not at the miter.

With all of the treads and risers assembled, I quickly level out any discrepancies with a smoothing plane. You could also use a sander here, but a sharp plane is faster, quieter, and less dusty. The final step is to run the treads across a router table that's still set up from rounding over the ends of the returns.

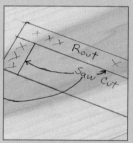

Building Housed-Stringer Stairs

Housed stringers are a step up from notched stringers. They're stronger, less likely to squeak, and once you've got the technique down, faster to build. In the simplest version (shown in this chapter), both stringers are mortised to accept the ends of the treads and risers. In chapter 5, I'll show you how to combine a housed stringer against a wall with a notched stringer on an open side for formal stairs.

One of the big advantages of housed-stringer stairs is that the finished product is that: finished. No building rough stairs and then trimming them out. That's the time savings these stairs offer—build them in the shop, and install them in one visit to the site. I prefer to build most of my main staircases this way, as well as commonly used utility stairs such as those going to basements or from the garage to the house. I generally make utility stairs from poplar or southern yellow pine because these species are inexpensive, strong, and readily available. Poplar is a better choice because it's more stable than southern yellow pine. Depending on the house, I may use poplar stringers and risers that will be painted and hardwood treads for main stairs. I've also built many that were all hardwood such as oak. These get heavy.

The Hows and Whys of Housed Stringers

Unlike notched-stringer stairs, housed-stringer stairs don't rely for support on heavy timber stringers underlying the treads and risers. Housed stringers are typically made from 5/4 x 10 or 5/4 x 12 stock. A router, guided by a shopmade jig, cuts mortises for the ends of the treads and risers into the stringers. The back sides of the mortises are cut at an angle, and glued wedges driven into the mortises lock the treads and risers in place.

How is it, you might wonder, that two 5/4 x 10 housed stringers are strong enough to support a stair that, were it a notched stringer, I'd use three 11½-in. x 1¾-in. LVLs for? It all has to do with how depth affects the strengths of materials. Depth, that is, the axis of a material that's parallel to its load, is a key determinant of the material's load-bearing capacity. Since notching a 2x12 or an LVL reduces its effective width to only 5 in. or so, it's pretty clear that a piece of 5/4 x 10, even with its thickness reduced to ¾ in. by the mortise, is the stronger piece of wood.

Wedges and glue lock the parts together. Housed-stringer stairs rely on tapered mortises that house the ends of the risers and treads. All parts are glued together, making exceptionally strong and squeak-free stairs.

Most housed stringers are made from 5/4 x 10. Never use thinner stock, as these are structural stringers. If there's a long, unsupported span, use 12-in. stock or 8/4 material instead.

There's no need to add a central notched stringer to housed-stringer stairs because each riser acts as a beam to carry its load to the stringers. Kept from flexing laterally, a 1x8 riser is quite strong. Consider I-joists, the engineered lumber that most floor systems are made from today. Their center web is only ½-in.-thick OSB, but it's kept from flexing by wider flanges top and bottom. Spaced 16 in. on center, 7½-in. I-joists can readily span 10 ft. Risers are usually spaced at most 11 in. apart, so there's no need for the central stringer, which would just add weight and potential squeaks.

Depth Determines Material Strength

Loading the same board in different ways demonstrates the importance of the depth of materials. This is why a full-depth 5/4 x 10 stringer can carry a greater load than a notched 2x12 stringer.

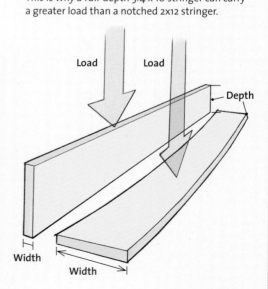

Open-riser option

I sometimes make housed stringers from heavier material. But that's almost always an aesthetic choice. For example, I've built housed-stringer stairs using 4x12 stringers and treads for timber-framed houses to create a rustic look. Such stairs often have open risers, which meet code because the treads are thick enough that the resulting space is less than 4 in. Since the risers are open, the mortises are visible. I do these a little differently. Few people would want to look at an exposed wedged mortise, so it's typical to cut the mortises to fit the treads exactly. Lag bolts through the stringer and into their end grain hold the treads in place.

Built in the shop

Unlike most notched-stringer stairs, I don't usually assemble housed-stringer stairs in place, unless they are extraordinarily heavy units or ones so long that I fear they'll fall out of my truck during delivery. In these cases, I make all their parts in my shop and transport them to the site. One big downside of on-site assembly is the mess. In the shop, stairs are assembled upside down on horses. It's easy to control glue drips. When they're assembled in place on-site, I have to work from below. Not only are the lower treads and risers uncomfortable to work on, but it's never long before I'm covered in glue drips.

Router-Cut Housed Stringers

There are other approaches to making housed stringers (see the sidebar at right), but this is the simplest I've found. The key to the process is a shop-made plywood jig and a pattern-cutting router bit. Pattern bits are straight bits with a top-mounted bearing that's the same

Housed Stringers of Old

I've seen housed-stringer stairs in Victorian houses, **built with the same wedged-mortise construction used today. The difference then, I suspect, is that brawny apprentices were used in place of a router to cut the mortises.**

Back in the day, each entire mortise—riser, tread, wedge angle, and nosing—would have been laid out on the stringers. Lines bisecting the tread and riser would have been drawn. Then, the mortise would have been roughed out using a brace and a series of ever larger bits to create the angle for the wedge. Cleanup would have been with a chisel, and the wedges must have been ripped by hand.

I suspect housed stringers were used only for the best class of work, as I've only ever seen them used in old houses for the main, formal stairs.

You can also make housed stringers using a straight, unguided bit in a router equipped with a guide bushing. I don't like this technique because the guide bushing spaces the bit 1/8 in. or more away from the guide, which introduces unnecessary complexity and inaccuracy.

diameter as the cutter. In use, the bearing rides along a template, in this case the stair jig, copying its shape exactly into the wood. The mortises extend out the bottom side of the stringers, allowing access to drive the wedges. This isn't the prettiest aspect of stairbuilding, but the bottoms of most stairs get drywalled or are in a utility area such as a basement.

When I build stair jigs, I always make them so that the riser and tread cutouts are about 3 in. to 4 in. longer than the actual risers and treads will be. Since the mortises extend

Making a mortising jig

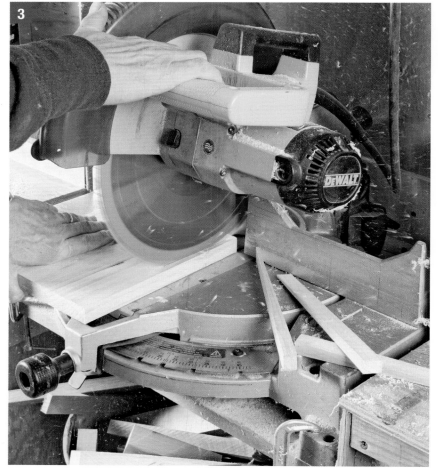

1 The jig must be at least 1 in. thick. Two layers of void-free ½-in. plywood underlayment glued together work well.

2 Begin layout by drawing a right angle, which represents the faces of the treads and risers.

3 Wedges are cut from scrap 1x stock and tapered to 4°. Crosscut the 1x to 10 in. long, then flip the miter saw to 2°. Working with the grain, trim the edge. Flip the board end for end and cut again, eyeballing the narrow end of the wedge at about ⅛ in.

4 Continue layout by holding the appropriate stock—tread or riser— and a wedge against the initial right angle. Draw a pencil line, which represents the back or bottom of the risers and treads.

5 Draw the nosing out past the original right angle. It should be the thickness of the treads and 1 in. to 1¼ in. beyond the face of the riser.

6 To guarantee straight cuts, clamp the jig down, and cut using a guide. Start the cuts with the saw's depth-control unlocked, the sole flat on the jig, and its edge tight to the guide. Lower the saw's motor and blade to full depth and make the cut.

7 Don't overcut with the circular saw. Instead, finish the corners using a jigsaw fitted with a new, sharp blade.

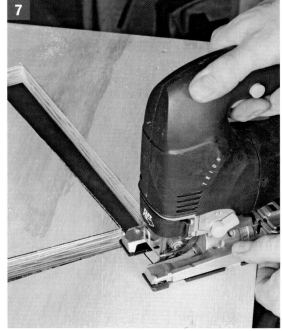

Plywood mortising jig guides a router. **Cut to the profile of the treads' and risers' faces with the back angled to create space for wedges, this shopmade jig works in conjunction with a bearing-guided pattern-routing bit. The bit's bearing follows the jig, exactly reproducing its shape as a mortise in the stringer.**

beyond the back of the stairs anyway, it's no problem for the jig to do so as well. And making the jig larger than the rise and run of the current set of stairs means it can be used for many others. The jig is infinitely adjustable within its range.

Making the mortising jig

I make the jig 1 in. thick by laminating together two pieces of ½-in. plywood (16 in. x 24 in. is a good size). I've not been able to find a pattern-routing bit that works well with a thinner jig. The cutters are too long, and if I make the jig from ¾-in. plywood, by the time the bit is lowered deep enough for the bearing to engage the jig, I'm making a deep, hogging cut that's hard on the router. Typically, I make stair mortises about ⅜-in.-deep, and the 1-in.-thick jig allows me to do this in two easy ³⁄₁₆-in.-deep passes.

After the glue sets up, lay out the jig as shown in the photos on pp. 62–63. The front of the riser and the top of the tread are simply a right angle drawn out with the aid of a framing square. The back layout lines are the thickness of a tread or a riser, plus a wedge, from this right angle. I make wedges about 10 in. long and at about 4° of taper.

Careful layout and cutting is crucial—any slips you make creating the jig will be mirrored in the stringers. It's important that the faces that will show on the stringer be clean, straight cuts. I use a circular saw and a guide for these cuts. I finish the tight cuts around the nosing overhang using a quality jigsaw and a sharp blade. For the first few jigs I made, I rounded the nosing cutout by first drilling it with a sharp bit of the same radius my treads would have. This was tricky because the hole had to be drilled exactly tangent to the top and bottom of the tread. One day I realized that if I simply radiused all treads at ⅜ in., the same radius as the ¾-in. router bit I mortised with, this little bit of fussiness would go away. Now I just cut the jig square, and the bit takes care of the radius.

The cuts for the backs of the risers and the bottoms of the treads don't have to be as precise. I cut them by eye, but if your hand on a circular saw isn't steady, use guides here as well.

Laying Out the Stringers

Layout for housed-stringer stairs is much the same as for notched-stringer stairs. You'll need the same tools—a framing square, stair gauges, and a pencil—as well as a keen eye for consistency. There are a couple of differences, however. First, notched-stringer stairs are laid out by aligning the square with the stringer's top, since that's where the cuts are made. Housed stringers are laid out by aligning the square

with their bottom, since that's where their mortises begin.

When laying out the stringer, pay attention to crowns and bows. Lumber is rarely perfectly straight, something carpenters learn to deal with. Crowns are arches across the width of a board, and bows are arches across its thickness. Make sure to lay out the stringers so that both crowns face upward. That way, they'll tend to straighten out with time and loading. Face the bows in opposite directions, either out or in. In this way, the bows cancel each other out and result in a straight set of stairs. I prefer to face the bows in because that helps during assembly by reducing the need to clamp the stringers

Lumber is rarely perfectly straight, **but most of the time its defects can be worked with. Sight the edge for crown and bow. Crowns always face up, and bows should face in.**

Bow and Crown

Wood is almost never straight, at least in the lengths used for stairs and carpentry. Carpenters figure out ways to work with the wood by orienting crowns upward like an arch, where loading will eventually straighten them. Bows are oriented to cancel each other out or so they're easy to correct with fasteners.

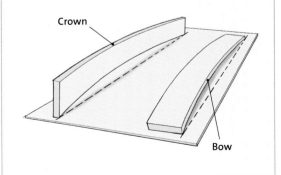

Crown

Bow

Lay Out Housed Stringers from a Secondary Line

Laid out from the stringer's bottom edge, the treads and risers would hang below the stringer. By laying out the stairs using a secondary line about 1 1/2 in. in from the stringer's bottom, the rear tread and riser intersections happen neatly inside the stringer.

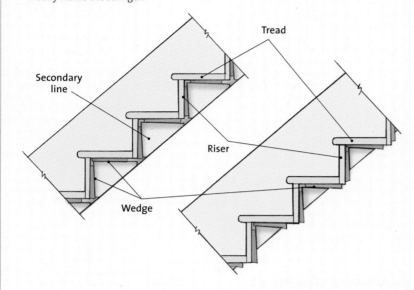

Tread

Secondary line

Riser

Wedge

Treads and risers don't meet at the edge of the stringer. The backs of the treads meet the bottoms of the risers at a line that's about 1 1/2 in. up from the bottom of the stringer. This depth fully encloses the treads and risers in the stringers for maximum support.

Rise and run dimensions ride the line; stair gauges ride the edge. Layout is a matter of sliding the square along the stringer and marking out the treads and risers so their intersections all occur on the layout line.

together to keep the risers and treads seated in their mortises. However, sometimes you have to face the bows out so that flaws in the face of the lumber don't show. (Even so, careful layout can often hide flaws such as knots behind the risers and treads, allowing the bows to face in.)

The second, critical difference between housed stringers and notched stringers is that the layout of notched stringers represents the back of the risers and the bottom of the treads. The front intersections of rise and run are congruent with the top edge of the stringer. Housed-stringer layout, on the other hand, represents the front of the risers and the tops of the treads and is done from the bottom edge of the stringer. Sort of. If you make the rear intersections of the faces of the treads and risers congruent with the bottom edge of the stringer, the backs of the treads and risers will protrude below the stringer. To avoid this, I set a combination square to strike a secondary line in about 1 1/2 in. from the bottom of the stringer. You can play with the location of this line if you want to show more or less of the stringer above the tread nosings.

Once you have the secondary line drawn, layout proceeds apace. Align the rise and run numbers on your framing square with this line, and set the stair gauges to ride on the bottom of the stringer. Unlike laying out a notched stringer, you don't reduce the bottom riser height by a tread thickness—that's because you're working from the top of the tread, not the bottom. The bottom riser itself will, of course, be narrower than all the others by a tread thickness. The level cut on the bottom of the stringer is simply an extension of what would have been the top of the next tread layout.

Making the bottom plumb cut

The bottom plumb cut should be thought out, since housed stringers continue beyond the bottom riser at least far enough to accommodate the nosing of the bottom tread. If that stringer meets another at a landing, for example, it's important to leave enough stringer so that it can be trimmed on-site for a neat corner. When stairs meet at a landing, make sure the plumb cuts on both the upper and lower flights are the same height. That way, the landing can be trimmed with a piece of 5/4 stock that matches these heights, and whatever molding is used to trim the stringers to the walls can continue around the landing uninterrupted.

Making the top plumb cut

The top plumb cut is another spot where it's easy to slip up. The layout at the top of the stair has to accommodate the landing tread. Most rabbeted landing treads match the tread thickness only where they overhang the top riser. When laying out for their mortises, I create an ear on the stringer that extends onto the landing or upper floor by the same distance as the landing tread. The bottom of this ear is a level cut. The top plumb cut rises from the end of that level cut.

I use the same technique for cutting the stringers to length at the top riser. Since the layout line represents the face of the riser, its plumb cut is the thickness of the riser behind the layout line. There won't be a wedge here, but the top riser fits in rabbets cut into the stringer using the router and template. To avoid making a wedge-shaped mortise, just make one pass with the router along the front edge of the riser part of the template. Cut out the ear with a circular saw, but don't overcut

Plan the intersections of flights at landings. **The top of this lower stringer was laid out to be long enough to mate with the upper stringer. The upper stringer was cut a bit long at the shop, then trimmed on-site. Trim will later blend the intersection of these stairs.**

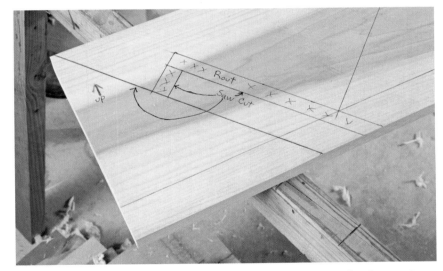

Layout for the top riser and landing tread looks different. **The landing tread (shown at left) is much shallower than the other treads, and neither it nor the top riser requires a tapered mortise. The three saw cuts shown will create a notch that will rest on the upper floor or landing.**

the intersection of the top riser's plumb cut and the landing tread's level cut. Finish the cut with a handsaw. Overcutting increases the chance of splitting the stringer at this point during installation.

Laying out the second stringer

It doesn't matter if you lay out the left or the right stringer first. What does matter is that after laying out the first one, you place it next to the second one so that their lower edges butt together. Using a framing square, transfer the locations of the rear tread and riser intersections from the laid-out stringer to the other one. Small, light hash marks on the secondary line are all that's needed. Their purpose is to maintain consistency and avoid accumulating error. When you lay out the second stringer, the rear tread and riser intersections should fall right on these marks. Notice I said, "should." As often as not, one or two will be off by a hair.

Unlike basic notched stringers, housed stringers come in lefts and rights. It's pretty easy to keep this straight with one stair, and it's just as easy to get it confused if you're building multiple stairs at a time. It's a good habit to mark each stringer for its location and for left or right.

Unless this is a big error—say, more than ⅛ in.—just split the difference. You might have to plane the back of a tread down by a couple of swipes at assembly, but no one will ever notice this small a difference. If you skip this step, however, it's easy to accumulate enough error by the end of a long stringer so that one is noticeably longer than the other. You might be able to assemble such stairs, but they won't be square, and they'll fight you every inch of the way at installation.

After laying out the frist stringer, ensure that the stringers mirror each other by transferring the locations of the tread/ riser intersections to the second stringer. A pencil mark across the second stringer's layout line shows where to align the framing square to prevent error from accumulating.

Mortising the Stringers

Now comes the fun part: routing out the stringer mortises. For this, you'll need a heavy-duty plunge router and a sharp bit. I cut my first sets of housed stringers using a fixed-base, 1½-horsepower Porter-Cable 690®. This venerable and reliable machine tried hard, but it quickly became evident that if I continued to use the 690 in this way, its days were numbered. And, the fixed base was an annoyance. I upgraded to a 2½-horsepower Bosch® plunge router, the biggest machine I could find locally at the time. Although it's seen a little repair-shop time, the Bosch continues to scream along.

To use the jig, simply align it with your layout marks and clamp. Where you start routing is critical: Always rout the left stringer from the bottom to the top and the right from top to bottom. This prevents blowouts and chips where the mortises intersect at the backs of the treads. Don't rout too deeply, either. You

Stair mortises make lots of dust and chips. A shopmade chip collector aligns with the stair jig and captures most of the chips and dust. It's held to the bench by a clamp that's hidden by the flexible duct.

Align the front of the jig with the layout lines. If you place the jig so that it just exposes the pencil lines, the router will remove them, saving cleanup. To keep the C-clamps holding the jig in place clear of the router, face the handles down.

Making Clean Cuts

When cutting a board at an angle with a circular saw, always cut "downhill" to the workpiece as shown below. This way, every wood fiber is supported by a longer one behind it, and the cut is nice and clean. The cut on the scrap piece will be chipped and fuzzy, but it's firewood anyway.

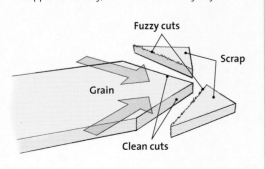

Fuzzy cuts

Scrap

Grain

Clean cuts

Dust Control

Mortising out a full set of stringers **creates an enormous amount of chips and dust. In my younger days, I didn't worry too much about dust control. I know now that wood dust is both an allergen and a carcinogen. I'm much more careful these days and have rigged my shop with dust control. A simple plenum catches the vast majority of the dust I create by mortising stringers. You could do the same thing with a quality vacuum set up to suck the dust from your router.**

won't save that much time, and you'll wear out your router and your bits prematurely. I usually make two passes at a cutting depth of ³⁄₁₆ in. in poplar or pine stringers. If I'm working in oak or some other dense wood, I'll make three passes ⅛ in. deep. Since the back of the landing tread and the top riser are both typically ¾ in. thick and no wedges are used here, keep these router cuts just one ¾-in. bit wide. Use the backsides of the resulting mortises as guides for sawing the stringer.

After both stringers are mortised, use a circular saw to make the plumb and level cuts. These are the finished stringers, so they'll show. Consequently, I cut going downhill on the grain and clean up the cuts using a sharp block plane. The area behind the top riser is completely cut away, and I finish the circular-saw cuts using a handsaw.

Finish the circular saw cuts with a handsaw. When routing for the top riser and the landing tread, just make one ³⁄₄-in.-wide mortise with no taper. Then, cut the notch using the back edge of the mortise as the guide for your saw.

Making smooth cuts. Any time wood is cut at an angle, one side of the kerf will go downhill on the grain and one side will go uphill. The uphill side will splinter and the downhill side will not. Always cut downhill on the piece you're keeping.

Rout in the Correct Direction to Avoid Chipping

The router should always move clockwise around the jig. Right stringers are routed top to bottom, and left stringers from bottom to top. This approach keeps the router under control and avoids visible blowouts caused by exiting one mortise into a part of another that will show.

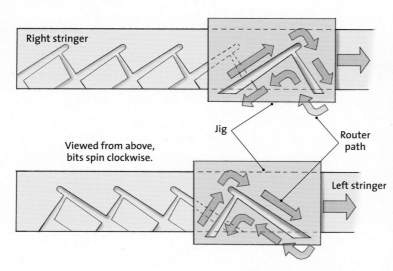

Right stringer

Jig

Router path

Viewed from above, bits spin clockwise.

Left stringer

Plane it clean. A couple of passes with a low-angle block plane clean the saw marks from the end grain. You could also sand them off, but a sharp plane is far faster and leaves a flat, crisp surface.

Cutting the Treads and Risers

I prefer not to wrestle long lengths of lumber through the tablesaw, so I cut the treads and risers to length before ripping them to width. Housed-stringer stairs should be 2 in. narrower than the distance between the framing, which provides enough maneuver room to ease installation of the stairs and leaves room for shims and drywall (it's always easiest to install the stairs before the drywall goes up).

Treads and risers are shorter than the stair is wide by the thickness of the combined stringer stock remaining after the mortise is cut. Using 5/4 stringers and ⅜-in.-deep mortises, that's usually about 1½ in. (¾ in. on each side). So, if the stairwell is 38 in. wide, the treads and risers would be 34½ in. wide (38 in. – 2 in. – 1½ in. = 34½ in.). This helps in avoiding material waste, too: 12-ft.-long-stock, if there aren't major end checks, can net four treads or risers; 16-ft.-long-stock allows for trimming the ends, with a smaller percentage of waste than, say, 10-ft. stock.

Your stair will only be as square as your treads and risers, so be careful as you cut them. A radial-arm saw or a large sliding compound-miter saw is ideal, but I've managed for years to get good, square cuts using a circular saw and a homemade shooting board.

Calculating widths isn't hard. The treads are simply the unit run plus the nosing overhang in width. The risers are mostly the unit rise in height. The exceptions are the top and bottom risers. The top riser's height is the unit rise plus the depth of the rabbet in the landing tread. The bottom riser is one tread thickness shorter than the unit rise.

A shooting board ensures a square cut from a circular saw. Figure the tread and riser length by deducting the thickness of the stock remaining behind the mortises in the stringers from the overall stair width.

Assembling the Parts

With all the parts cut and mortised, two wedges for each tread and riser at hand, and a full glue bottle (I use yellow carpenter's glue—Elmer's® or Titebond® I, II, or III), it's time to assemble the stairs. I build them upside down on sawhorses. Be sure the horses are of consistent height and on a flat, if not level, floor. Otherwise, you'll build a twist into the stairs that you surely don't want.

Installing the treads

Start with the bottom tread, first checking it for bow (any bow should face upward). With the tread in its mortises, use bar clamps top and bottom to clamp the stringers together. Don't overtighten either clamp or you'll angle the stringers in or out. Check that the tread length fits the mortise correctly, then glue up a wedge and drive it home. I don't glue the top side of

Installing the treads

1 Begin with the top and bottom treads. Clamping the stringers around these treads helps to square up the stair. Once these treads are wedged and glued into place, no further clamping is likely to be needed (as long as you faced the bows of the stringers inward).

2 After seating the tread, glue up the top, bottom, and inside of the wedge, and hammer it home. The first couple of hammer blows may wedge the tread backward in the mortise, so tap the tread forward as you go.

3 The landing tread gets no wedge. Instead, run a small bead of glue in the mortise, drill a countersunk pilot hole in each side of the landing tread, and screw it home.

4 Trim the excess from the tread wedges. Often, these will drive beyond the riser mortise, and a chisel makes fast work of the excess. Any wedge left hanging behind the treads should be trimmed with a handsaw.

the tread. That always results in glue showing where it's not wanted, and the wedged-and-glued connection below has never earned me a callback for a squeaky stair. Wipe off any excess glue, then move to the topmost full tread and install it the same way. Installing these two treads should square the stringers to each other, which you can verify by measuring the diagonals. If the diagonals are off, it shouldn't be by much unless you've made a major layout error. Pulling the stringers a little bit in opposite directions will equalize the diagonals, then you can proceed with hammering home the wedges.

Once all the treads are in, install the landing tread, using a bead of glue and a couple of 1⅝-in. screws (I prefer coated deck screws for this application). Trim off the wedges' excess length with a handsaw, and you're ready to install the risers.

Installing the risers

To best fit the risers to the treads, first check them for crown and bow. Crown should face up to match any bow the treads have. Bow in the risers isn't crucial unless the stair is exceptionally wide, but I do try to face it backward. That way, the pocket screws I use to affix riser to tread tend to straighten out the riser. There is a distinct sequence to applying glue to the risers. To minimize drips, apply glue to the top of the riser first, then, when that's in place, glue the back of the tread. As with the treads, use glued wedges to pin the risers in place.

I use a Kreg Jig and pocket screws to attach the risers to the treads above. For this, I usually use 1¼-in. screws intended for pocket-hole joinery, which are available from several sources. These screws have heads with flat bases that exert great pressure on the bottom

Work with the curve of the wood. **Sight the riser to find the direction of its bow, then apply a bead of glue to that edge.**

Work quickly to avoid drips, **after turning the riser over and sliding it home in its mortise. As soon as the riser bottoms on the tread below (below, that is, until you right the stair; then it's the tread above), the glue bead will smear and reduce the chance of a drip.**

Glue the back of the tread. **Pull the riser away from the tread before it, and run a bead of glue along the tread's upper edge. Gravity will pull the glue down, but don't wait too long before seating the riser.**

of the pilot hole, drawing together the riser and tread. Cone-head screws, such as deck screws or drywall screws, tend to wedge wood fibers apart, so I don't use them for this purpose. Pocket screws come in both coarse and fine thread. Coarse-thread screws excel at entering soft woods such as pine and poplar, and their deep threads engage without stripping, as fine-thread screws might. Fine-thread screws thread more easily into hardwoods and are less likely to shear from the torque of being driven.

Another Way to Reinforce the Riser-to-Tread Joint

I built many stairs before owning a pocket-screw jig, and my old method of reinforcing this joint worked fine, too. I wouldn't glue the top of the riser to the tread per se. Rather, I'd finish wedging the risers in place, then reinforce this joint in the back with rippings from the risers or treads that measured approximately 3/4 in. square and were nearly as long as the entire joint. Pour on the glue to both the back of the riser and the bottom of the tread, "squooge" the ripping into the glue, and drive three or four screws at an angle through the ripping and into the tread. Angling the screws pulls the ripping firmly into the corner, making for a tremendously sound glue joint.

Pocket screws pull the riser to the tread, in effect clamping the glue joint. Pocket screws go into angled pilot holes drilled with a special jig and drill bit. Alternatively, a cleat can be screwed and glued in place to reinforce this joint.

Wedge the riser home, again being generous with the glue. As with the treads, wedging can move the riser out of its mortise. Keep an eye open, and tap the riser downward occasionally.

Use a long driver to set pocket screws. Short drivers lead the screw gun into the riser, putting the driver at an angle where it's likely to strip the screw head before setting it. Pocket screws have a wide head and a smooth upper shank designed to pull joints tight.

While you've got the drill out for the pocket screws, use the same bit to drill pilot holes in the backs of the risers. **Readily available deck screws attach the riser to the back of the tread.**

Pocket screws join the landing tread and riser. **After carefully spreading a bead of glue in the landing tread's rabbet, the stringers' mortises, and on the back of the tread, place the top riser and screw it home.**

While I've got the pocket-screw drill bit out, I drill pilot holes in the bottom of the riser, too. Deck screws or cabinet screws pull the riser to the back of the tread before it, clamping the glue joint. Although they look like drywall screws, deck screws and cabinet screws are superior products for this application. Drywall screws have little shear strength and could break.

The top riser, of course, gets no wedges but is held in place with screws and glue. Once the top riser is in, go back and cut off all the wedges, flip over the stair, and check for errant glue. The final step is to cut and nail in place the cove molding below the nosings.

Installing the Stair

The hardest part about installing stairs is moving them. They're heavy, and you'll need help. Most basement stairs, for example, can be installed by two strong people, but more is better.

Cove trims below the nosings. **It's hard to measure for cove molding (sometimes called scotia molding) with a tape. Fitting overlapping sticks to the stringers and clamping them together yields a numberless and exact inside measurement.**

Installing basement stairs

For basement stairs, the sequence is generally something like this. With the stair on its side partway overhanging the stairwell, one person goes down and one stays up. The person above slides the stair just beyond its balance point and gently lowers the end until the person below can reach and control it. The stair is lowered until it's more or less vertical in the well and clear of the upper doorway. At this point, its' ready to be rotated to its proper orientation in the stairwell.

The person below can now begin to pull the bottom of the stair forward, and the person above helps as best as he or she can, mainly

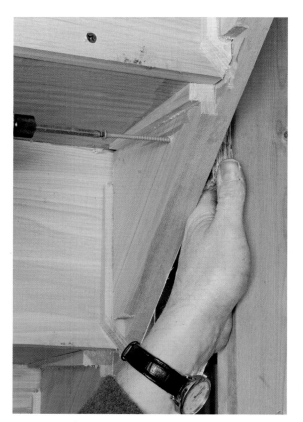

Fasten stringers to the framing from below. At installation, shim between the stringer and the studs, and drive long screws from the back of the stairs.

by acting as a guide. When the stair reaches its proper position, it should just drop into place on the ledger. At that point, the landing tread and stringers are carrying all the weight. Center the stair in the opening, and drive a couple of 10d finish nails through the top riser to temporarily secure the stair. The person above can now walk down the stair, as long as the person below is there bracing the bottom to prevent it from kicking out.

Working from below, shim between the stair and each stud, driving 3½-in. deck screws or landscaping screws through the shims. Keep the shims below the tops of the stringers to allow drywall to slide in. Secured in this way, the stair isn't going to move.

Installing stairs between floors

For stairs to be installed from below, the procedure differs. Safety first: If this stair is above a lower stair, install the lower stair first. That way, you won't fall as far. Then, install some walking planks over the lower stairwell. The best practice is to lay framing lumber such as 2x10s on the flat in every stud bay and top that with a layer of plywood so you don't step through the gaps. (If you're working on a slab floor, ignore the latter instruction.)

With the opening protected, stand the stair up vertically, just behind where it'll land on the lower floor. Get one or two helpers on the upper floor, and have them help to rotate the stair into its proper orientation and lower it so that the top end of the stair is resting on the upper floor. Then, with a helper, lift the bottom end of the stair and carry it to its landing place. Set it down, check side-to-side clearances as before, and shim and secure.

Building a Combination Stair

A combination stair has one housed stringer, usually against a wall, and one notched stringer on the other side, which is typically open to a foyer. Sometimes, there will be a wall partway enclosing the stairwell, and the notched stringer goes only as far as the end of this wall and transitions to a housed stringer for the rest of the flight. Unlike the open stairs discussed in chapter 3, there is normally no underlying rough stringer. The notched stringer is typically made from 5/4 x 12 and is supported by a wall below. Without this wall, a notched piece of 5/4 x 12 lacks the strength to be a structural stringer on a stair longer than two or three risers.

Combining housed and notched stringers is by far my favorite approach to building formal stairs. It takes the work out of the chaos of a job site and places it in the controlled environment of a shop. Working on stairs in a house that's under construction is akin to road repairs. Somebody always wants to get through, and you're always an inconvenience to them. The reverse is true as well—like a road crew, you have to make allowances to keep the traffic flowing, which cuts into the time you can devote to just doing your job.

Measuring and Planning

Most of the measuring techniques discussed in earlier chapters apply to combination stairs as well. You'll use the same methods for determining the overall rise, locating the landing height, and dealing with out-of-level floors.

Combination stairs use both notched and housed stringers. **The trick to these shop-built stairs, which are commonly used for main stairs, is knowing which assembly techniques to borrow from which type of stringer.**

What's a little different is figuring out the width of the stairs.

Notched-stringer stairs must be supported by a wall that is built just to the inside of the stringer. Its top plate affixes directly to the rear tread and riser intersection. This wall's width must be figured into the stair's overall width. I usually build this wall immediately after installing the stairs. (I rarely allow anyone else to do it, either. My concern is leaving an unsupported stair on the job. If someone uses the stairs before the wall is built, there's a chance the stairs could break. It seems unwise to assume that liability on a promise that the supporting wall will be built "soon.")

How I build the wall depends on the configuration of the stairs. If they go straight up to a landing or the next floor, I can usually build a standard 2x4 wall that aligns with the opening for the basement stair below. (Of course, if you're working from a slab floor, this isn't a consideration.) In this case, the wall starts at the edge of the stairwell below, so its thickness is that of the studs, 3½ in., plus that of the wall finish, usually ½-in. drywall. I add ¼ in. to that to give the drywall crew a little room to fit drywall behind the stringer. That small gap will be covered by trim.

For stairs whose notched stringer is interrupted by a full-height wall partly enclosing the stairwell, the supporting wall under the stair has to be narrower than the full-height wall. Otherwise, the supporting wall would hang over the cellar stairwell. Fortunately, typical stair dimensions are such that building the wall under the notched stringer with studs placed on the flat makes the stair meet the full-height wall just as you'd probably prefer.

Calculating Stair Width

Unless they're designed with heavy-duty engineered-lumber stringers, most notched stringers must be supported by a wall, which must be factored in when determining the width of the stair. Two common situations arise. The first is when the notched stringer simply intersects a landing or the upper floor (top drawing). The second is when the stairwell is open to one or both sides at the bottom and enclosed by walls at the top (bottom drawing).

Straight Run into a Landing

This is the simplest situation, as a standard 2x4 wall can be built below the stairs. The notched stringer overhangs the wall, with space left for drywall plus a little wiggle room.

Tread length =
well width + stud width + notched stringer width + tread overhang – housed stringer thickness after mortising

Riser length =
tread length – tread overhang

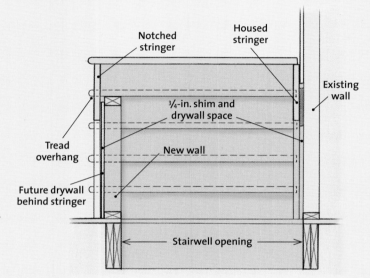

Partly Open Stair Intersects the End of an Enclosing Wall

This is a common situation. The wall below the stair must be built with the studs on the flat, and the drywall will butt to the stringer rather than run behind it. Otherwise, the top tread on the notched stringer will protrude beyond the enclosing wall.

Tread length =
well width + stud width + notched stringer width + tread overhang – housed stringer thickness after mortising– 3/4-in. shim and drywall space

Riser length =
tread length – tread overhang

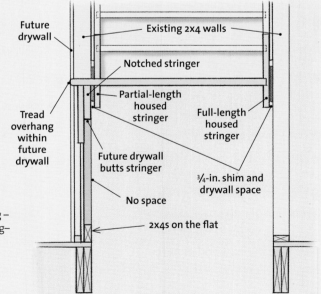

Laying Out the Stringers

As with basic housed-stringer stairs, I start by evaluating the stringer stock. There's a difference, though. With housed stringers that are mirror images of each other, it's easy enough to oppose the stringers' bows so that they cancel each other out and make a straight stair (see pp. 65–66). Because the stringers of a combination stair differ, I use the straightest stock I can find for both of them.

Laying out the notched stringer

Lay out the face of the notched stringer first, using a framing square and stair gauges. This layout is no different from laying out any notched-stringer stair. Make sure the crown faces up, and if the lumber has defects such as knots or pitch pockets, place these in the notches where they'll be cut out.

Finished notched stringers lay out like any notched stringer. A framing square with stair gauges set at the rise and run lays out the stringer. The fact that the riser cuts will be miters doesn't affect the layout.

Ending the Top of a Notched Stringer

When a stair terminates at a landing or upper floor whose edge is parallel to the stringer, it's common to tie the two together visually with a trim board that runs below the landing tread. The stringer can run long, meeting the trim board at an angle (option 1). Or it can be plumb-cut directly below the back of the riser (option 2), just as it would be if the stair intersected the landing or upper floor at 90° (option 3).

With option 1, the top riser meets the stringer at a miter. With options 2 and 3, the top riser can be square cut. The end grain of the riser will be hidden by the newel. With all three options, the thickness of the top riser must be added to the unit run of the top tread.

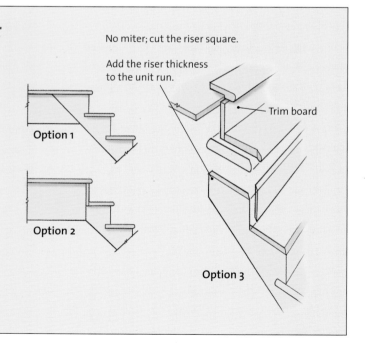

No miter; cut the riser square.

Add the riser thickness to the unit run.

Option 1

Option 2

Trim board

Option 3

Adjust the bottom and top riser layouts. A scrap of tread material held on the framing square eliminates math in deducting one tread thickness from the bottom riser cut. The top plumb cut is one riser thickness behind the back of the tread layout and is easily marked using a riser scrap.

One choice you'll need to make is how far to extend the top of the notched stringer. If the landing or balcony continues the line of the notched stringer, you can let the stringer run beyond the top riser and trim it off flush with the subfloor. Done this way, any skirtboard on the landing will intersect the stringer at an oblique angle. If the stair hits the landing at a right angle, then it's a simple matter of plumb-cutting the stringer at the back of the top riser. Doing so leaves a fragile bit of mitered stringer whose sole purpose would be to return the mitered end of the top riser to the wall. I almost never bother with this return, preferring to simply extend the cut for the top tread all the way. There's little point to returning the top tread, since this detail gets buried behind the landing newel.

An easy way to lay out the riser adjustments at the top and bottom of the stair is to use scraps of the appropriate stock in conjunction with the framing square. At the top, the unit run of the last tread has to extend the thickness of the top riser farther. Just align the square as if you were going to lay out any riser, but hold a scrap of riser stock behind the square and pencil along that. The bottom riser cut on a notched stringer is the thickness of a tread short. Again, hold the square to make the level cut back from the standard riser's height, but hold a piece of tread stock against the square to mark the stringer. These approaches avoid measuring and ensure exactness.

Laying out and routing the housed stringer

Once the notched stringer is laid out, start on the housed stringer. As in the previous chapter (see p. 66), run a pencil line down the length of the stringer about 1½ in. up from the bottom edge. Then, use the laid-out notched stringer and a framing square to make reference marks along the pencil line to keep the stringers consistent. Once the layout is done, rout the housed stringer in exactly the same way as in chapter 4, using the plywood jig clamped to the stringer and a pattern-routing bit chucked into a plunge router.

Cutting the notched stringer

Cutting the notched stringer isn't rocket science—it's just cutting to a line. I make the miter cuts first, setting the blade on my circular saw to 45° using a Speed Square or a gauge block cut on an accurate miter saw. I don't trust the bevel gauge on my circular saw. With only a few high-end exceptions, circular saws are intended as framing carpenters' tools. They can be used as finish tools, but they must be

Finish cuts need accurate setup. **A circular saw is designed for rough work, so its own bevel settings can't be trusted for miter-cutting stringers. A scrap of wood cut to 45° is precise enough.**

> **When cutting out notched stringers** with a circular saw, you'll need a saw with the blade on the right for right-hand stringers. For left-hand notched stringers, use a left-bladed saw.

set up with more accurate guides. Alternatively, you can make this cut using a sliding compound-miter saw. I finally acquired one of these beasts about halfway through writing this book, and my circular saw sees a lot less use now.

As with the finished stringer in chapter 3, you must miter this stringer as perfectly as possible. I use a shopmade shooting board (see p. 55) to do this and finish the cuts with a handsaw. After making the miter cuts, set the saw back to square and make the cuts for the treads. Exactness isn't as critical here, as these joints are hidden with molding. Still, just to avoid error, I use the shooting board for these cuts, too. Any cleanup is done with a sharp paring chisel. Unlike the stringer in chapter 3, the notch cutouts aren't scrap. Save them for glue blocks to be used during assembly.

Installing the Treads and Risers

The treads and risers are made as in chapter 3, except that the depth of the mortise in the housed stringer must be added to the length of the treads and risers. Putting things together is much different, however. I developed my method of assembly with an eye to two things: (a) using glue in a way that prevents the stair from squeaking over the long haul, and (b) preventing glue drips during assembly. These goals are at odds with each other, but I reconcile them by gluing blocking in place behind the stair parts themselves.

Shopmade guide ensures straight cuts. A straight-edged piece of 1/2-in. plywood glued to a 1/4-in.-thick base guides the saw. Make the base overwide, and the first cut made with the guide trims the base to exactly where the blade cuts. Separate guides are needed for each saw and each bevel angle.

Left-blade saw cuts left miters. Because of the direction of blade tilt, a left-blade saw can't cut right stringer miters and vice versa.

Hand tools still have a place. Notches in stringers should never be over-cut, so circular saw cuts must be finished with a handsaw. A sharp paring chisel cleans up what's left.

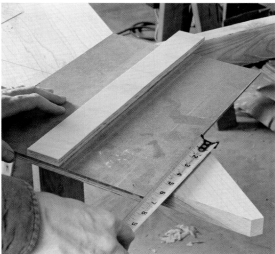

Match the bottom plumb cut to the width of the landing base molding. In this case, the landing will be trimmed with 5/4 x 6, meaning that the plumb cut should be 5 1/2 in. in height for a perfectly continuous line between the stringer and the base.

Installing the risers

1 Risers go into the housed stringer first. Use a scrap of tread material to keep the riser properly spaced as you drive its wedge. Be careful to keep the riser fully seated in its mortise.

2 With the wedge driven to refusal, trim it flush with the bottom of the riser.

3 The top riser doesn't get a wedge. Instead, carefully apply a bead of glue to the front of the rabbet and drive in three 1⅝-in. screws to secure the riser to the housed stringer.

4 You'll need both glue and finish nails to ensure a long-lasting miter.

5 Drive 1½-in. finish nails through the stringer into the ends of each riser.

6 Sand down any high points of the miter while the glue is still wet to fill imperfections with a dust-and-glue slurry.

Ready for treads. After dry-fitting to check tread length, apply glue to the back edge of the tread. Spread out the glue into a uniform film to minimize squeeze-out.

Installing the risers

Place the housed stringer mortise-side up on a flat surface such as a concrete slab. A low bench can work as well, but since combination stairs are assembled on their side, much of the work takes place 3 ft. or more in the air. Wedge and glue the risers into place, placing a scrap of tread stock in the tread mortises to prevent the risers from intruding into the tread mortise. Set the risers fully in their mortises and trim the backs of the wedges so they don't protrude beyond the bottom of the riser. Otherwise, these wedges will interfere when you're driving the wedges for the treads. The top riser, of course, is screwed into place.

With all of the risers installed, next comes the notched stringer. It's useful to have a helper to hold one end of long notched stringers, but a stand of the appropriate height will work as well (and you won't owe the stand a beer at the end of the day). Glue and nail the risers to the stringer, using yellow glue and 1½-in. finish nails. It's much easier to use a nail gun here than to hand-nail. Start at one end, and carefully align the miters on the riser and stringer. Secure the first riser with several nails, then move to the other end of the stair and repeat. After the top and bottom risers are secure, nail off the rest of the risers.

Installing the treads

The next step, installing the treads, is the most awkward part of the process. Before spreading any glue, check each tread's fit. These treads have mitered returns that extend past the back of the tread, as was shown in chapter 3 (see pp. 52–57). Does the return fit tightly to the stringer with the tread in place? If not, pull the tread and trim enough from the mortise end so that the return seats against the stringer. Now, with the tread out of the stair, run a bead of glue along its back edge. Then, insert the tread

Careful now—that's wet glue. First, place the tread nose into the mortise, then seat the tread in the mortise and against the face of the riser. Be careful to keep the returned side of the tread flat on its notch to avoid getting glue where it's not wanted.

Angled block gives clamp purchase. Cut at the same pitch as the stair and with sandpaper glued to the edge that contacts the stringer, a wood block gives the clamp something to pull against. The clamp secures the tread while it's affixed from below.

into the mortise at an angle, with the nosing entering the mortise first. With the tread flat on its notch, rotate it in the mortise and against the riser behind. Use a bar clamp and a triangular block to ensure that the tread is seated on the notch.

Moving to the back side of the stair, drive a glued wedge into the mortise, then drill pilot holes and screw the riser to the tread with 1⅝-in. deck screws. Finish up by using a pocket-hole jig to drill holes in the stringer to connect it to the tread. Be careful to keep these screws away from the points where holes for balusters will later be drilled. Keep the front screws back 1½ in. or so from the riser below, and set the back screw about the same distance from the back of the tread.

Locking the treads in place

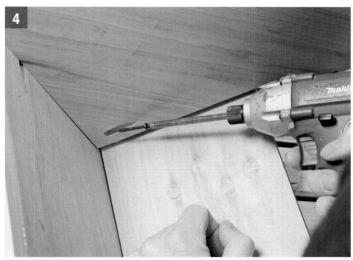

1 After confirming that the tread is fully seated in the mortise, prepare a wedge to lock the tread in place.

2 Screw the riser and tread together. Placed in countersunk holes to prevent splitting, these screws mainly serve to clamp the glue joint between riser and tread.

3 Use a pocket-hole jig to drill holes into the notched stringer to connect it to the tread.

4 No glue is used in this joint, as it's hard to prevent drips that show. The pocket screws serve only to hold the tread in place until a glue block is installed. Take care not to screw where you'll later drill for balusters.

Adding blocking

At this point, no glue has been used between the notched stringer and the tread or between the lower riser and the tread. That's because I find it impossible to prevent drips when gluing these surfaces directly together. This is where blocking comes into play.

Although I usually use a pocket-screw jig to join the risers to the treads, as shown in chapter 4 (see p. 73–75), this connection can also be made with cleats. Some stair shops glue and staple in several short, triangular rippings, but I prefer a longer glue joint for strength. After ripping the riser stock to width, I'm always blessed with a pile of roughly ¾-in.-square strips (which is why I always order 1x10 stock for the risers, even

when 1x8 would be large enough). For use, these strips need to be shorter than the space between the housed stringer and the glue blocks on the notched side. I gang-cut the strips to length on a miter saw, smear two edges with glue, and screw them home. I use enough 1⅝-in. screws toed into the tread to pull the strips into full contact with the riser and the tread.

Earlier, I said not to throw away the triangular cutouts from the notched stringer. These are used to make the glue blocks for the connection between the tread and the notched stringer. Since one face of each block is mitered, they need to be trimmed to provide a glue surface. Use a miter saw to cut off the angled edge and then to square the other leg

Glue strips join the front edge of the tread and the riser below. **To avoid drips, don't use any glue directly between the tread and riser here. Instead, glue and screw wood strips from behind. Cut the strips about 2 in. short to provide clearance on the notched stringer side for the glue blocks shown next.**

Stringer cutouts become glue blocks. **Trim off the mitered end first, then rotate the block 90° to trim the other leg of the triangle at a right angle.**

Use glue, and lots of it. **The glue blocks reinforce the miter joint, as well as join the tread to the notched stringer (top left). There's little risk of glue getting to a finished face from here, so use it freely. Angle the pilot holes so that the screws pull the block toward the tread (bottom left).**

Rosin paper protects the treads. **Affix wood strips from scrap with 3/4-in. staples to hold the rosin paper in place. One-quarter-inch plywood or Masonite (textured side up for safety) offers better but more costly protection.**

of the triangle with that. Affix these blocks to the back of the notched stringer, the riser, and the tread using screws and copious amounts of glue. This is a messy process.

With combination stairs, I generally don't install the landing tread at this time. It will have to be removed and trimmed to fit the landing newel when I'm installing the railing, so there's no point to installing it now. The final steps are to trim the wedges (and usually the glue block at the bottom tread because it protrudes below the bottom riser), install the cove molding (see pp. 51–52), and add a protective layer of rosin paper.

You may be concerned that the joints between the block, the tread, and the riser are end-grain glue joints that won't hold, but don't worry. Because of the angled cuts, they're not quite end or edge grain, and the joints do in fact hold well.

Join the stringers before fixing them to the framing. **With the stairs in place, first line up the joints between stringers and secure them. Then shim and screw the stairs in place.**

For studs on the flat, use a top plate that hangs below the stringer. **Screwed to the back of the notched stringer, this top plate provides great nailing for drywall and trim and makes installing the studs easier.**

Installing the Stair

Although there are ways to build a freestanding notched-stringer stair (see the sidebar on the facing page), a combination stair as described here requires a supporting wall below the notched stringer. The first step is to wrestle the stairs into place, checking for level as well as verifying top and bottom riser height. If two stringers meet at a landing, join them with 8d finish nails.

Building a support wall

Shim and screw the housed stringer to the wall as described in chapter 4 (see p. 76), and then you're ready to build the support wall. As with any wall, start with the top and bottom plates. There are only two methods—either the studs go in edgewise, as with most walls, or on the flat. The inside of the wall should be flush with the stairwell below to provide an uninterrupted plane for the drywall.

If the wall is conventionally built, with the studs on edge, simply screw the top plate to the bottoms of the risers, being careful to use screws short enough not to go all the way to the face of the stairs. If the studs are on the flat, affix the top plate by screwing through it into the stringer, again being careful not to screw all the way through. With studs on the flat, I skip the bottom plate and nail the studs to the subfloor, aligning them on a chalkline. I'll fill the spaces between the bottoms of the studs with blocking. This approach is a little easier than either toenailing a 2x4 on edge to the floor or ripping one in half to serve as a plate.

A laser plumb bob is an ideal tool for transferring the layout from the bottom plate to the top, but a level or a traditional string-powered

Mark the studs in place. **With the stud held to layout marks on the floor and plumbed with a level, one pencil swipe marks the length and the cutting angle.**

Sometimes, you skip the bottom plate. **When the studs go in on the flat, it's easier to toenail them to the floor and fill between later with blocking.**

plumb bob works as well. The easiest approach to marking the studs to length is to rough-cut them just slightly longer than needed, hold them in place, and mark them there. It's easiest to work from the bottom of the stair up, as this leaves you some working room. Be careful to install the studs snugly but not so snugly as to wedge the bottom of the stair upward. This is an easy mistake to make.

Freestanding Stairs

Rarely, I'm called on to build a freestanding stair. If it's freestanding on both sides, it's usually easiest to build the stair on-site. If it's freestanding on one side only, then I build it pretty much as discussed earlier in this chapter, except that I reinforce the stair by adding engineered-lumber substringers, such as were covered in chapter 2 (see the sidebar on p. 25). After the stringers are in place, I install shorter glue blocks at the tread and riser intersections.

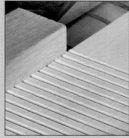

Adding a Bullnose Step

A staple of Colonial and Colonial Revival homes, a bullnose step is the lowest step in the stair. Bullnose steps extend in a half-circle beyond the notched stringer on one side of a stair that runs along a wall or on both sides of a freestanding stair. Unlike most treads and risers, bullnose steps are assembled as a unit consisting of the tread and riser and are installed on the stringers as one piece. Bullnose steps can be purchased from railing suppliers and even some home centers. The disadvantage to buying a prebuilt bullnose step is that unless you can allow the unit run of the bullnose to dictate that of the rest of the stair, you're facing an awkward trimming job where the bullnose tread returns against the stringer. The unit rise presents the same problem. Rather than fight these battles, I usually make my own bullnose steps.

When making my own starting steps, I build in a mortise for the volute newel as well. (A volute newel is a post that springs from a bullnose step to support a railing volute.) It's pretty easy to glue up the bending form, which also serves to anchor the newel on most bullnose steps, with this mortise. If I'm going to use a 3-in.-wide newel, I'll make the mortise 2½ in. or so square

A bullnose step is mortised for a starting newel. **This makes a stout attachment that's a strong argument for making your own bullnose starting steps.**

A shopmade mortise is dead square to the tread. **When the tread is installed level, the mortise in the starting step is automatically plumb and so too will be the newel.**

and trim the newel's bottom as a tenon. This leaves a shoulder that braces the connection a bit and hides any minor imperfections in the tread mortise. Sound like a lot of trouble for little value? I don't think so. Because it's built in shop conditions, it's a fairly simple matter to make the mortise square to the tread. Install the tread level in the house, and the volute newel is automatically plumb. And a 2½-in. tenon provides a lot of glue surface and a lot of beef. It's a sturdy attachment.

Designing the Bullnose

The thing that drives the design of any bullnose step is the volute, the spiraling section that starts the railing. The volute must center on the bullnose, and the balusters need to land on the tread so that they don't overhang the roundover. Every volute I've ever used came with a paper template intended for use in locating the volute on a starting step and in

laying out the baluster spacing. It's also where I look for the shape of the bullnose step. Volute templates not only show the approximate shape of the volute and several possible baluster layouts, but they also provide the information needed to create the perimeter of the starting step.

Larger volutes can seem to call for a bullnose that's deeper than the common treads of a stair (at least on a Connecticut stair, where the code allows a 9-in. unit run). This doesn't mean that the bullnose starting step can differ in depth from the rest of the stair. What happens is that the radius of the bullnose increases to the back, while the portion of the tread that gets walked on remains the same depth as all the other treads. In other words, the back part of the bullnose that returns against the stringer is deeper than the rest of the starting step. The radius of the bullnose is tangent to the front of the tread. It makes its half-circle and then

Attaching a Newel on a Store-Bought Bullnose Step

Store-bought starting steps are generally used with a volute newel that mounts by way of a long, integral dowel. For these newels to be plumb, you have to drill a plumb hole through the tread and the solid section that acts as the form for bending the riser. I do it using a long (12-in. or so) 1/4-in. twist bit, often sold as a "bellhanger's bit," to drill a pilot hole. Two plywood squares guide this bit, but it still rarely drills a perfectly plumb hole.

I follow up the twist bit with a spade bit of the appropriate size (usually 1 1/4 in.) that's mounted on an extension. If the pilot hole is straight and plumb, it all works out well, and on the test fit, the newel slips snugly into place, plumb as a falling stone.

All too often, though, I end up chamfering the bottom of the newel's dowel to provide clearance so it will sit plumb. This can be a time-consuming affair of test-fitting, shaving, test-fitting, and more shaving. And at that point, getting the newel plumb is only part of the battle. After planing or rasping away enough material from the dowel so that the newel will be plumb, the dowel is, of course, loose in the hole.

Assuming the stair is over a basement, that's not a big deal. I'll have drilled the hole all the way through the subfloor, anyway, and most newel dowels are long enough to stick through the subfloor. I spread glue in the hole and on the dowel, and set the newel. Then, working from below, I hammer shims into the gap. Give this a moment's thought first. If the dowel isn't sitting true, and you drive the shim on the wrong side, the glue will probably set nearly instantly, and the newel will be permanently out of plumb. The procedure is essentially the same if the stair is over a crawlspace, except you have to duck-walk over to where you see the glue dripping. If the stair is on a slab foundation, your options are limited.

A thick dowel secures this newel. Usually 1 1/4 in. in diameter, the dowel is glued into a hole drilled plumb into the tread.

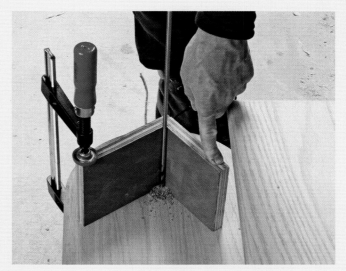

A plywood jig guides a pilot bit. Two pieces of plywood screwed square together and clamped to a level step keep a 1/4-in. bellhanger's bit plumb.

A spade bit follows the pilot hole. A quick-release extension provides the length needed for the long dowel.

Establishing the perimeter of the bullnose

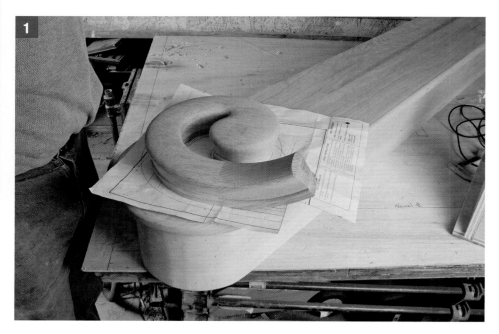

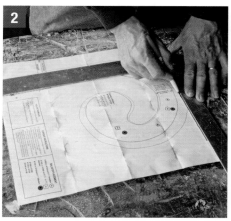

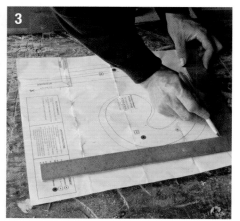

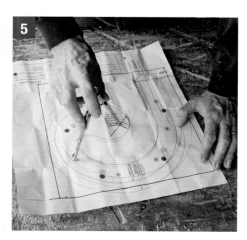

1 Pick the railing volute first. The size and shape of the bullnose depends on the railing volute going above. Volutes come with a paper template, useful for laying out the balustrade later and the step now.

2 Align the square with the stringer line on the template, and mark a line representing the front of the tread. Look to the baluster locations shown on the template to check that this width will work.

3 Find the end of the bullnose by squaring off the tread front. Note the distance between the front of the tread and the line of the volute, and repeat this on the end. The return at the back of the tread simply continues the nosing overhang back to the stringer.

4 Mark the center of the bullnose radius on the template. This center will not be the same as the volute's center; it's found at the intersection of a pair of 45° angles coming from the squared off end of the tread.

5 Scribe the bullnose radius on the template. This is more of a check than anything, and it allows you to ensure that the volute fits the bullnose and that the balusters have sufficient landing room.

Prepare to glue an extension to the tread for the return. **The edge of the stringer starts the return. Mark the stock for a biscuit joint. Overlay the template on the tread, to check that the biscuits won't be exposed when cutting the radius.**

Biscuits play two roles. **Biscuits not only help to keep the faces flush but also to limit side-to-side movement. That eases aligning the return on the mark representing the stringer.**

Clamping from below aids in visibility, **keeps the stock from being glued to the bench, and eliminates noticeable black stains on the oak from the reaction of the water-based glue and the steel clamp bars.**

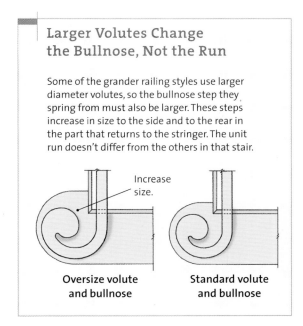

Larger Volutes Change the Bullnose, Not the Run

Some of the grander railing styles use larger diameter volutes, so the bullnose step they spring from must also be larger. These steps increase in size to the side and to the rear in the part that returns to the stringer. The unit run doesn't differ from the others in that stair.

Increase size.

Oversize volute and bullnose

Standard volute and bullnose

returns with a short straight section to the stringer.

Since the volute pattern determines the shape of the bullnose tread, everything else springs from it, including the width of the tread itself. Even the smallest bullnose is deeper than the other treads where it extends beyond the stringer. At the very least, it's the depth of a standard tread plus the depth of the overhang on the back side of the bullnose. Larger volutes call for bullnose treads that are not only wider but also extend farther beyond the stringer.

Making the Tread

One way to make the tread is to use a wider piece of stock and notch it for the tread. To the frugal Yankee in me, this seems wasteful, as well as calling for an awkward finish cut at the notch. Instead, I work with a longer piece of the same tread stock as on the rest of the stair and simply glue on a section to serve as the back of the bullnose. If I make a nice square

cut and glue this extension on just right, I've made the notch. I make sure the tread runs a little long from the end of the applied piece and trim the far end. If it's a double bullnose step, then I take care to glue on both extensions so they're spaced at exactly the outside width of the finish stringers.

I use biscuits to ensure a good alignment of the tread and the extension, but I take care to keep the biscuit well inboard from where I'll saw the curve into the tread. Few things look

Solid Stock for Bullnose Treads

Manufactured tread stock is often made from a core of seemingly random hardwood scraps glued together, with a solid nosing glued to the front and thick veneers glued to the top and bottom. Most of the time, these treads are fine and make good use of marginal lumber. The ends are either buried in closed stringers or returns are applied. The end grain, which shows off the veneer layers and the assembly of miscellaneous hardwood strips, is covered. Not so on bullnose steps. The round end pretty much requires a show of end grain. Here, I always use solid stock.

more amateurish than a biscuit joint exposed by a finish cut. After the glue is set, I scrape off any that I can, then plane and sand the joint flush.

Although the paper template guides the manufacture of the bullnose step, I don't lay out the radius cut directly from it. I do lay out the radius on the template, as well as the location of the mortise, but I use that only as a model for the actual layout. The template locates the far edge of the bullnose and shows where the newel's center will be (which is not

the same point as the center of the bullnose's radius). But the template doesn't locate the center of the bullnose radius, and trying to trace a radius from a paper template is too inexact. I use a compass with its radius set to half the width of the bullnose tread at the extension. Finding the center is a matter of measuring in the distance of the radius from the end point of the tread and from the front or back of the tread, using a ruler or the set compass. From this point, I lay out the radius on the end of the tread stock. The resulting half-circle should be tangent to the front and back of the tread. It's good practice to do this layout on the bottom of the tread, to ease layout of the riser. More on that later.

Mortising the tread

Mortising the tread can be intimidating, but it's not that big a deal. After laying out the mortise on top of the tread, drill a series of starter holes. Then, using a quality jigsaw (this is really no place for a $39.99 hardware store tool), cut close to the lines. Use a freshly sharpened paring chisel for final cleanup.

Draw the newel on the template. Use a framing square to mark the shape of the newel around the volute's center. This can be transferred to the tread by measuring or by laying the template atop the tread and marking the four corners with an awl.

A compass marks the tread for cutting. Note again that the center of the bullnose and the center of the newel are not the same. The radius should be tangent to the front and the back of the tread.

Mortising the tread

1 | Using measurements or points taken from the template, lay out the newel's mortise on the tread. Align the square on the front of the tread to ensure that the sides of the mortise are square to the tread.

2 | Be careful to keep the starter holes inboard of the layout lines. A ½-in. or ⅝-in. spade bit is sufficient to get most jigsaw blades started.

3 | Here's where a quality jigsaw and sharp blades pay off. It's an important cut, and inexpensive jigsaws don't hold the blade reliably square to the work. Leave the line, and take your time.

4 | Clean up the mortise with a sharp paring chisel. A square-edged block clamped to the tread guides the chisel quite precisely.

Jigsaw the radius of the bullnose. **Stay to the outside of the line, and cut from each side to the apex of the radius. This cut direction eliminates splintering on the side you'll keep, making the line easier to see and the cut smoother.**

Block the tread off the bench, and clamp. **This allows the bench to keep the belt sander square to the tread. Carefully sneak up on the line, and remove any flat spots. Sand with the grain, just as you jigsawed with the grain, for the smoothest finish.**

Bullnosing the tread

Cutting out the circle is almost an afterthought because most of the work comes in laying it out. I've used a bandsaw but actually prefer a jigsaw for this task. A 4-ft.- or 5-ft.-long oak tread is an unwieldy thing to drag across a bandsaw table. Make the cut starting at the front, and stop halfway through. Start over again at the back, meeting up with the first cut at the apex of the end grain. By cutting this way, with the grain or uphill, the cut is always supported by longer grain behind and splintering is greatly reduced. Be careful not to cut inside of the radius, as you'll create a flat spot that's difficult to get rid of. Leaving the line isn't a big deal because you'll take care of that in the next step.

Fairing the radius is quickly done with 100-grit paper on a belt sander (coarser grits can be too hard to control). This is an important step: There should be no saw marks when you're done and no flat spots either. Run your fingers around the curve—you can feel defects that you can't see. As with sawing, sanding

Round over the tread with a ⅜-in.-radius bearing-guided bit, **routing with the grain to yield the smoothest cut. This means climb-cutting half the radius on each side, but the bit is small enough that caution is all that's needed.**

uphill yields a smoother finish. Any irregularities that you leave will be reflected in the roundover.

After brushing or vacuuming the sanding grit from the tread, it's time to rout the roundover. I use a bearing-guided ⅜-in. radius bit. Again, I pay attention to the grain direction on the radius, starting in both the front and the back. Caution: This means that you'll always be climb-cutting on half the bullnose.

Making a Half-Round Riser

Shaping the tread is the easy part. The riser is the part that stops a lot of people from making bullnose steps because it has to be bent. Bending wood can be scary, but it's really not that hard.

With the tread upside down, adjust the compass to the bullnose radius minus the nosing overhang. Swinging this arc on the tread shows where the outside of the riser will end up. Close the compass by the thickness of the riser, and scribe another half-circle on the tread. That's the future location of the inside of the riser and the outside of the bending form that's next on the to-build list.

The bending form will be as high as the bottom riser or a little less. Its width is the diameter of the last half-circle drawn on the bottom of the tread. Its length runs from the outside of the finish stringer, along a line drawn square across the tread from the notch, to the far point of the radius. I make it up by stacking and gluing solid stock, often culled from the scrap bin since none of the pieces are very long.

Now, if you're content to use a doweled starting newel (see the sidebar on p. 95), you can skip the parts that follow about making each layer of the form out of several pieces. Jump right ahead to the part about bandsawing the form.

Making the form

To make a form with an integral mortise, first lay out the mortise on the paper template, being careful to center it on the newel and to

Climb-Cutting

Climb-cutting means that you're moving the router so its bit engages the stock from the opposite direction the router is moving. Remember that bit from high school physics about every action having an equal and opposite reaction? Climb-cutting exemplifies this dictum, and the router will want to fly out of control. Controlling it isn't a problem with a relatively small bit like the one I use here, assuming that it's chucked in a fairly heavy router like my Porter-Cable 1 ½-hp machine. I would hesitate to use such a bit in, say, a laminate trimmer. Be aware.

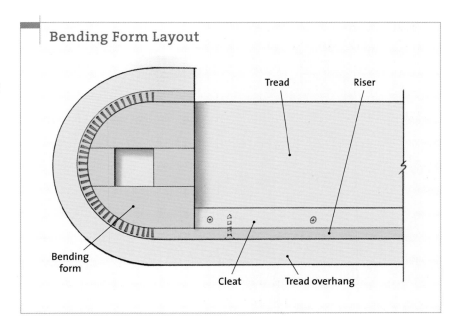

Bending Form Layout

Tread

Riser

Bending form

Cleat

Tread overhang

Lay out the mortise on the paper template, being careful to center it on the newel and to keep it square to everything else.

Making the bending form

1 Assemble the layers of the bending form from stock cut to fit around the mortise and inside the riser. Splines align the pieces, so kerf the stock before cutting to length. Chisel out spline stock intruding into the mortise later.

2 Use a scrap of stock cut to the size of the newel's tenon to align the parts of each layer at glue-up. Cut the parts of the form long for later trimming so that you don't have to fuss when the glue is running.

3 With the tenon scrap in the tread mortise, lay one form layer in place. This will be marked and cut as a template for the succeeding layers.

4 The tenon scrap not only locates the template layer but also provides a place to center the compass. Although the bullnose and the newel centers differ, the bullnose center falls within the mortise.

5 While it's on the tread, mark the edge of the stringer on the first form layer. Square-cut it on a miter saw.

6 Bandsaw the form layers. Since the first layer will serve as a template for the others, cut carefully, then sand to the line. For the others, just try to split the line. Final sanding happens after the form is glued up.

7 Alignment is critical when making a mortise. A guide the size of the tenon ensures accuracy. Wrap the guide in plastic to prevent glue from sticking to it.

8 Use clamps to bring the layers together and force out excess glue. Although the form doesn't show at all, wiping off the excess glue keeps it from gumming up the sandpaper in later steps.

9 Sand the form fair. Even with the aligning tenon, the layers will not be perfect. A belt sander and 80-grit paper soon right that. Keep the sander flat on the form to avoid rounding over the top and bottom.

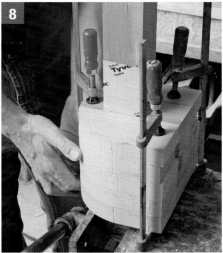

keep it square to everything else. Each layer of the form will consist of four pieces that surround the mortise. To find the widths of the front and back pieces, measure from the mortise layout to the outside of the bending form layout at the front and back of the step. Because the newel doesn't center on the bullnose's center point, the parts of the form before and behind the newel are usually of different widths. Rip enough of each so that when cut to length and stacked, there's enough material to add up to the thickness you need for the form (about 6 in.). The pieces to the outside and inside of the mortise are the same width as the mortise.

Since each layer consists of four pieces and the complete form consists of a stack of these layers glued together, precision in assembly is necessary. The four pieces that make up each layer have to be glued up so they're almost flush. The quickest way to accomplish this is to spline the joints. This is a simple matter of setting up a tablesaw to kerf the stock edges (one edge of the outer pieces and both edges of the center ones). Do this before cutting the stock to length so that you're running longer, safer pieces over the tablesaw. I kerf this stock noticeably off center to make its top and bottom obvious. Flip the center stock end-for-end to ensure that the off-center kerfs are off center closest to the same face. I cut the splines to fit snugly but so that hand assembly is possible.

Laying the various pieces of stock in place on the template reveals their rough lengths; cut them up on a chopsaw. Before glue-up, make up a chunk of square stock the same size as the tenon on the newel will be. Wrapped in plastic—Saran® Wrap is just dandy—to keep it from being glued in place, this post serves as a form for the form pieces. Apply glue and

splines to these pieces, and assemble them around the post. Clamped for five minutes or so, a layer can then be removed from the post and set aside to cure.

After the glue sets, scrape off the worst of it, locate the center of the bullnose radius (often that's in the mortise, and you have to temporarily fill the hole with a slice from the form post to give the compass somewhere to swing from), and draw the radius on a template layer. While it's in place on the tread, mark this template layer where it hits the outside of the stringer. Square-cut it there using a miter saw, then go over to the bandsaw (a jigsaw would work, too, but the bandsaw is easier to use on

Join the form to the tread. **Using the scrap tenon to align the form and the tread, glue and clamp them together. Before the glue sets, use a square to check that the form is aligned with where the stringer will be. If it's out a little, a few hammer blows will put it right.**

these smaller pieces) to cut the radius. Try to split the line, and rasp away any flats. I use this first layer as a pattern to lay out succeeding layers for bandsawing, indexing them together with the form post to ensure consistency and tracing the shape to the uncut layer with a pencil.

With the radius cut on all the layers, look them over quickly to be sure their surfaces are flush.

To complete the assembly of the form, place an even film of glue on each layer, then stack them on the post and clamp. Leave the post in place until the glue has fully set. At that point, unclamp the form, scrape off as much glue as possible, and fair it with a belt sander and 80-grit paper. Check for square regularly as you're sanding. Once the form is faired with the belt sander, scrape all the glue chunks off the top. Then spread on some new glue, and using the post as a guide, place the form on the bottom of the tread. Clamp it in place, being careful that the clamps don't squeeze the form outside of the layout lines.

Preparing the tread for the riser

The final step in preparing the tread for the riser is to install a cleat along the tread that serves to locate and support the straight section of the riser. Strike a line along the length of the tread representing the back of the riser, and screw this cleat in place just behind the line. If the bullnose is to mortise into a housed stringer, keep this cleat 1 in. or so back from the end of the tread so as not to interfere with the joint. The location of this cleat determines the riser's position, so position it carefully, particularly if the bullnose tread is to be part of a housed-stringer stair. You've got to be sure the tread and riser will fit the mortised stringer.

Mark the rest of the tread for a cleat to support the riser. **Set a combination square flush with the face of the form, and run the mark to the end of the tread. This mark represents the back of the riser and the front of the cleat.**

Screw the cleat to the tread. **While it's possible to pocket-screw the riser to the tread, there's already too much going on in the coming glue-up. This cleat eases assembly and ensures that the riser ends up properly placed.**

Preparing the Riser for Bending

The radius of the curved riser is that of the bullnose less the nosing overhang. I've bent risers in two ways, both requiring the support of a half-round form that becomes a permanent part of the bullnose tread. I prefer the approach discussed below, kerfing the riser stock, but you can also thin down the part of the riser to be bent and wrap that around a form, as shown in the drawing on p. 106.

The easiest way to bend the riser material is to repeatedly kerf its back using a radial-arm

Hollow-Back Riser

An alternative approach to bending a riser is to hollow out the back, leaving only a $1/8$-in. thickness of wood to go around the bending form. Leave a section of unhollowed stock several inches long at the end. A dado blade is the best way to hollow out the riser, followed up by smoothing with a plane. You have to measure the perimeter of the bending form with a tape measure to determine the length of stock to be hollowed and add about $3/4$ in. to that to have the necessary play to bring the unhollowed section past the form.

Making the bend in this way goes smoothly if you've gotten the form nearly perfect. Liberally coat the hollowed section with glue, butt the riser to the form, and pull the thin wood around. To tighten the riser to the form, drive opposing stair wedges into the $3/4$ in. of extra space in the hollow.

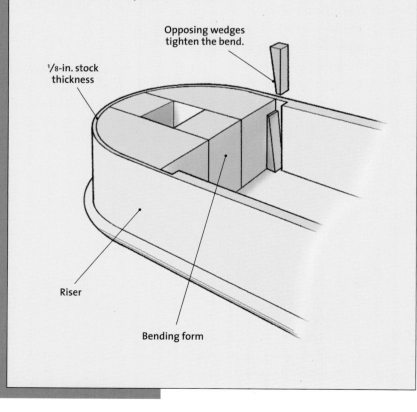

Opposing wedges tighten the bend.

$1/8$-in. stock thickness

Riser

Bending form

saw, compound-miter saw, circular saw, or tablesaw, leaving about $1/8$ in. of wood below the kerfs. It's crucial that these kerfs be spaced properly (more on this below) and that the kerfs be of a consistent depth or the wood won't bend well. You'll end up with flat spots if the spacing varies, and the wood will want to curve up or down if the kerfs' depth tapers.

When using a radial-arm saw, miter saw, or tablesaw to cut the 40 to 50 kerfs needed to bend around a typical bullnose step, be careful to maintain consistent down pressure on the workpiece. With a radial-arm or miter saw, if the work rises off the table, the kerf will be too deep and may even penetrate the face of the board. On a tablesaw, you've got the opposite worry. I prefer either other saw over the tablesaw for cutting all of these kerfs. It takes intense concentration to space the kerfs correctly and to keep the workpiece tight to the tablesaw's miter gauge. It's just easier (and probably safer) when you can see the sawblade interacting with the wood.

The spacing and number of kerfs necessary to bend a piece of wood vary with the radius. I've found that a kerf every $3/8$ in. (assuming a kerf that measures $1/8$ in. and leaves $1/4$ in. between kerfs) works quite well with poplar and pine risers. For an oak or other hardwood riser, $1/4$-in. spacing works better. Those spacings will allow a bend of a $33/4$-in. radius, typical for a 9-in.-run Connecticut stair (9 in. between the faces of the riser at the front and back of the bullnose, less the combined thickness of the riser front and back yields a form diameter of $71/2$ in.; divide that in half to find the radius of $33/4$ in.). These spacings are generous for larger runs.

If you want to figure the exact spacing needed for a perfect radius bend, there is a way to do it. Cut one kerf into the board, leaving ⅛ in. of material. Clamp the board to a flat surface, and mark its edge the distance of the required radius (3¾ in. in the case of this stair) from the kerf. Raise the unclamped end of the board until the kerf closes, and block the board in this position. Measure the distance between the flat surface and the face of the board now. That dimension is the minimum space between kerfs and should yield a perfect bend where all of the kerfs close against each other. Err on the side of making the kerfs a little too close, though. Too close, and the wood will still bend successfully. Too far apart, and you've just ruined a board.

Laying out the kerfs

Lay out the kerfs on the bottom riser, starting at a point that allows the riser to run a little past the end of the tread. An inch or so is all that's needed, although longer isn't a problem. The kerfs should begin about an inch before the curve of the form and extend a similar amount past the curve on the back of the form.

Sometimes, even though you've picked the right wood, prepared the form correctly, and kerfed the riser just right, things go terribly awry during the bend and the wood cracks or peels apart. Wood is a natural material that's not entirely predictable. A wise stairbuilder has a spare piece of stock on hand for this eventuality.

Straight Grain for Bending

Successful bending depends on having straight-grained wood. If there's any grain runout to speak of, it's likely that the face of the riser will split out during the bend. In all species that I've worked with, quartersawn boards bend best, with the least likelihood of splitting. This is particularly important with red oak. You don't necessarily need to search out lumber sold as quartersawn, either. Most lumberyards and hardwood dealers won't object if, as part of an order for a whole stair, you pick through the piles looking for one quartersawn board. You can tell quartersawn lumber by looking at the end grain. In a quartersawn board, the growth rings will be perpendicular to the board's face.

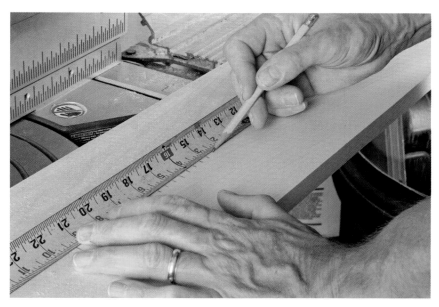

Consistent kerfs make for a smooth bend. Lay out all the kerfs on the stock to ensure consistency. In most instances, kerfs spaced ¼ in. apart, and leaving ⅛ in. or so of stock, will allow an easy bend.

Guide block speeds kerfing. **After the first kerf, slip a block between the fence and the stock. Align the blade on the layout mark for the second cut, and align the block on the first kerf. Clamp the block to serve as a fast-reading benchmark for the rest of the cuts.**

Cut away. **Most sliding miter saws can be adjusted to cut dados. Keep consistent down pressure on the stock so that it doesn't rise up into the blade, penetrating the face and ruining the riser.**

I've used two methods to find the length of the kerfed section. It can be done mathematically by multiplying π by the radius, or simply by wrapping a tape measure around the form. I favor the latter method. Make sure to let the riser run a bit past the back of the form, too. It's easy to cut to size later and pretty much impossible to do beforehand.

With the limits of the kerfed area marked, all you have to do is lay out the spacing. To space the cuts, begin by cutting the first kerf on a layout mark. Then, place a square-cut piece of scrap between the riser and the saw's fence. You'll clamp this in place in a minute. First, though, align the saw to cut the second kerf and then align the scrap wood with the already cut kerf. When you're happy with these alignments, clamp the scrap wood in place. Now for each cut, just align the previous kerf on the scrap, and cut away.

Bending the Riser

Here's where the rubber hits the road, one of the few moments in stairbuilding, or in carpentry in general, where you don't know for certain that the next 10 minutes will go as you've planned. It's a little stressful.

Earlier on, the form was glued and clamped to the tread. Make sure that all of the glue squeeze-out from this operation has been removed. Rest the tread on a series of bar clamps in preparation for clamping the riser to the cleat. Have ready half a dozen cauls a little taller than the riser to aid in clamping the riser around the form, and have open several deep-throated clamps that can slip into the mortise to secure the cauls.

The big moment: bending the riser

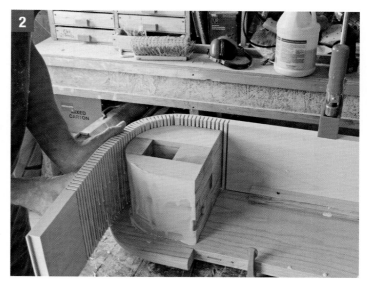

1 With glue dripping everywhere, anchor the riser by clamping it to the cleat. Don't skimp on the glue (and don't forget the glue between the riser and the cleat). Note that the form is shorter than the riser, which eases fitting the step to uneven floors.

2 Carefully swing the riser around, smoothing it to the form as you go, then hold it in place with one hand. With the other, you'll place cauls and clamps.

3 The mortise is a great anchor for clamping. Keep the cauls up off the tread, so they don't end up glued there. Start clamping at the front of the step and work backward. This moves any kinks along, instead of clamping them permanently in place.

4 Let the glue set up for 24 hours before unclamping, then saw off the excess riser flush with the back of the form. Finish by carefully hand-sanding the curved part of the riser; the stock is thin, and sanding through it will ruin your day.

Now, start spreading the glue. Coat the face of the cleat on the tread, and liberally coat the entire surface of the riser that will contact the form. Stand the riser in place, and clamp it to the cleat. Start bending the riser around the form, adding a clamp and a caul every 3 in. or 4 in. all the way around. That's the glue-up. Wipe off the glue that's dripped and be prepared to sand. Don't sweat the glue on the underside of the tread too much. That's not a place where many people look.

I like to allow yellow glue a full 24 hours to set up on a bullnose before unclamping. I hand-sand the curved riser with 100-grit paper on a block, and I take it easy. There's only 1/8 in. of material there, and it's possible to sand through. Don't ask how I know this.

To Cove or Not to Cove?

It's possible to make a cove molding to fit the joint on the curved section of the bullnose by using a bandsaw to cut an outer curve on some 4/4 stock that's twice the width of the cove wider than the riser. This curve is coved out with a router, and then the inside is bandsawn to match the outer radius of the riser.

But a curved cove is a detail I rarely bother with. It's time-consuming and no one to my knowledge has ever complained about its absence. In fact, you almost have to stand on your head to tell if there's a cove under a bullnose step, and few people undertake such acrobatics. If it salves the purist in you to make a curved cove, have at it. Life is short, though.

Assuming the tread is at the proper length, mark a cut line on the riser square up from the tread's end and cut it flush with a handsaw. The riser on the back side of the form should be trimmed even with the end of the nosing overhang, as both will butt to the finished stringer.

Attaching the Bullnose

Attaching the bullnose is really no big deal. If one end is to fit into a housed stringer, and you've been careful, it should just slide in, ready for wedges and glue. I run the notched stringer under the tread, just as anywhere else on the stair. Pocket screws secure it. A couple of screws through the stringer and into the returned riser should create a tight joint here.

If you're adding a bullnose to a site-built notched-stringer stair, you'll have to notch the points of the stringers to make clearance for the cleat backing up the bottom riser. The upper riser attaches to the tread before assembly, just as elsewhere on the stair. The lower part of the finished stringer just slides in behind the returned rise. You do have to shorten the bottom run by the thickness of the front riser as there's no miter joint here. Since there's probably no clearance to screw the bullnose tread to the rough stairs from below, be particularly generous with the construction adhesive. After setting the bullnose in place, stand on it to ensure good contact, and shoot a couple of 2½-in. finish nails into the stringers.

Winders

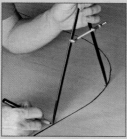

Let's get a couple of things straight first. Winders are not circular stairs, where the treads twine around a pole from one floor to the next (that one's a spiral stair). Nor are they curved stairs, whose laminated stringers carry the treads and risers in an arc to the second floor. While both of these types of stair share some geometry with winders, they are beyond the scope of this book.

Winders are stairs that turn corners, typically with a run of straight stairs above and/or below them. The turn is most often a 90° corner, although it can be 180°. Straight stairs can turn corners, too—with landings. In essence, instead of using a landing, which really is just one big tread, to turn a corner, winders squeeze several risers and treads into the same floor space. Our forbears knew this, as evidenced by how often one finds winding stairs in the tiny houses necessity forced them to build. Some of these older stairs are astoundingly steep, with winders that taper to nothing and are only 6 in. or so deep at the walk line. Stairs such as these suggest that the upper floors were the province of the young and the spry and that folks over 30 slept downstairs.

Winders climb rapidly at corners. **Making the most of space, winders pack multiple risers into the same space as a landing would take.**

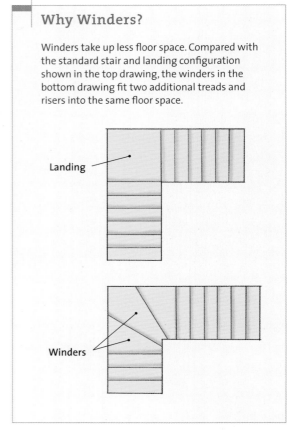

Why Winders?

Winders take up less floor space. Compared with the standard stair and landing configuration shown in the top drawing, the winders in the bottom drawing fit two additional treads and risers into the same floor space.

Landing

Winders

Winders and the Code

Modern winders are safer and more comfortable than those old ankle-breakers, but it's a little harder to build them. As recently as the 1980s, some codes had minimum tread-depth requirements for the walk line of winders but allowed the insides to taper to zero. Today's codes require a minimum tread width of 6 in. at the narrowest point. These old-style winders were far easier to build and shortened the run even more than is allowed today. All you had to do was build a landing and, essentially, stack treads and risers that were angled boxes made from framing lumber atop it. Straight runs of stairs meeting these boxes at the top and bottom completed the staircase.

The walk line is, to state the obvious, where most people walk on stairs. For code purposes, it's defined as being 12 in. from the inside ends (the narrow ends) of the treads.

A version of that approach can be taken today, but it loses some advantage because of headroom issues if there's a stairway below. The old way, with the landing forming the foundation of the so-called "platform" winders, took up no more run than any other landing. But because of today's 6-in. minimum tread requirement, that landing has to extend around the corner, intruding into the headspace of any stair descending below. If you're building on a slab or a crawlspace foundation, there's no problem. But if the winders stack atop a basement stair, odds are it won't be possible to maintain the required headroom.

Stringers for winders are usually straight, although they can curve. They can be open or closed. Typically, there will be a newel post at the inner corner of a winder, which can be structural—that is, it supports the stair.

Winders can be built without a newel here, but doing so necessitates a custom railing fitting that not only turns at the proper radius but also drops at the same slope as that particular stair. You don't see these much on new work, but they were more common in years past when carpenters made the rails and fittings on a custom basis as a matter of course.

Winder Stringers

The winders in this chapter have housed stringers on each side. The straight runs at the top and bottom are standard, although the inside stringer of the finished stair will be doubled in thickness for appearance. The riser heights of winders don't change, but the tread mortises on the wall stringer differ in length. The angles of each winding tread increase farther from 90° until the corner tread, and then they decrease back to 90° at the upper run of common treads and risers. Four of the five treads are the same at the inside. The fifth differs only because it's at the corner.

That's how most winders go, including the ones shown here. It's possible and code-legal to be more free-form. Code dictates a uniform tread width at the walk line of 12 in. Here's the wiggle room. Building codes specify only a minimum 6-in. tread width at the narrowest point. The code doesn't specify a uniform tread width here. By increasing it, you can build a stair resembling a chambered nautilus if that's what feeds your soul. Of course, you'd have to get the building inspector to sign off on it, and that's another matter.

The stringers at the winding part of the stairs change pitch—that is, the wall stringer is at a shallower pitch because the treads are wider, but the risers are the same height. The inside stringer is at a steeper pitch, and the

Winders and the Code

Gone are the days when the treads of winding stairs could taper down to nothing at the inside. International Residential Code–compliant winders have minimum tread-depth requirements at the inside, as well as at the walk line. Building inspectors usually have some latitude to grant exceptions for stairs being retrofit in an existing house.

The minimum tread depth of housed-stringer stairs ends up being the unit run where the treads and stringer intersect. It's more complicated with notched-stringer stairs because the minimum tread width is measured beyond the stringer, and it's an academic exercise to calculate the unit run without doing a full-scale drawing.

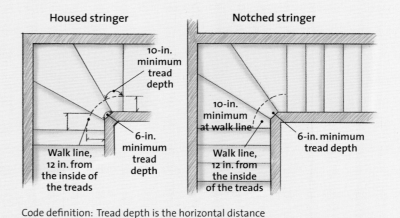

Code definition: Tread depth is the horizontal distance between the nosings of adjacent treads.

walk line should be at the same pitch as the straight runs. The change in pitch means that you have to pay particular attention when laying out the stringers. For example, because the pitch of the common steps is steeper than that of the winders at the wall stringer, fitting both commons and winders on one piece of stock may be impossible. The commons quickly drop off the bottom of a stringer laid out at the pitch of the wall winders.

In this stair, there was only one common tread at the bottom of the stair. Both the

Same stair, different pitches. Although the riser height doesn't vary, the tread depth changes dramatically from one side of the winder to the other. That makes the inner stringer far steeper than the outer stringer.

wall stringers and the inner stringers could accommodate that. Likewise, the top inner stringer was laid out at the pitch of the upper commons, and the stock was deep enough to accommodate the steep drop of the winders. The top wall stringer was the exception. It had to be joined to work with both the commons and the winders.

Sound complicated? It can be, if you try to think of it all at once. Like so many aspects of stairbuilding, the trick is to wrap your head around one component at a time. Eventually, it all comes together holistically.

Unlike curved stairs, the stringers used for winders are made from standard, straight lumber. Because the treads change width, the rise/run ratio and thus the pitch change as well. Depending on where and how much the pitch changes, it may be necessary to use a two-piece stringer.

Laying Out the Winder Stair

To figure out winders, first create a full-scale plan-view drawing of them. I do this on a sheet of ¼-in. plywood, or two if that's what it takes. I then project elevations of the winding stringers to figure out their depth and how they'll join.

Making a full-scale drawing

Drawing out winders full size allows you to figure out the sizes and the angles and eliminates guesswork when making the treads and risers. Because the stairwell length combines with the width of the lower set of stairs to limit the overall run of the upper stairs, tread layout always begins at the top and progresses downward. For housed stringers, the layout lines you'll draw represent the fronts of the risers. The treads overhang them by the same amount as on the straight runs of the stair.

Draw in the stringers. Draw both edges of all stringers, as well as a line on each representing the ⅜-in. depth of the tread mortise. Extend the plywood as far as it seems you'll need to accommodate the bottom run of stairs. At this point, how far they extend is unknown, so simply project two lines representing the outsides of the stringers.

Draw the perimeter of the stair on a sheet of plywood, **using the stairwell dimensions and the desired width of the stairs as starting points.**

Draw in the upper treads. With both stringers drawn, start filling in with the treads (just the unit run; don't include the nosings yet). Start from the lowest winder and work down (photo 1 on p. 116). Don't forget to add the thickness of the top riser. The unit run is likely to be code minimum because if you had room for bigger treads, odds are you wouldn't need the winders. Stop with the upper treads when you're 1 ft. or so away from the corner.

What you do next will depend on whether you're building a housed-stringer winder, as shown here, or a notched-stringer winder. Most codes specify that the narrow end of winder treads must be a minimum of 6 in. For housed-stringer stairs, that happens at the inside edge of the stringer, and that's where all of the walk-line measurements are taken from.

For notched-stringer stairs, that 6 in. happens beyond the stringer at the tread return. So in addition to drawing the stringer on the inside, you'll have to draw a line representing the overhang of the tread returns. That's where the walk-line measurements for notched-stringer stairs originate.

Draw in the winder treads. Measuring from the appropriate location, draw a walk line on both the upper and lower flights, 12 in. from the narrow end of the treads (photo 2 on p. 116). Set a compass to 12 in., then, pivoting at the point where the two lines representing the edges of the narrow ends of the treads meet, strike an arc going around the corner. That arc should be tangent with both walk lines (photo 3 on p. 116).

Moving to the lower flight, draw a line square to the stringers representing the back of the top common tread (a common tread is one that's not a winder tread). The back of this tread is the same distance down from the corner as is the front of the upper flight's first common tread. These two lines are the beginning and end of the winders.

To lay out the narrow ends of the winding treads, start by setting a pair of dividers to 6 in. (photo 4 on p. 116). Beginning at either of the common tread lines, step off the inner tread widths along the appropriate line (the inside of housed stringers and along the tread return for notched stringers). Unless you're lucky, this won't fall evenly. Open the dividers a little, and step it off again. Repeat this process until you hit on a dimension close to 6 in. that steps off evenly, then mark those spots on the plywood (photo 5 on p. 116).

Laying out the treads along the walk line follows a similar procedure (photo 6 on p. 116). The starting width of the dividers would be the unit run of the common treads, which is most likely the same as the code minimum unit run. Starting at either the upper or lower common tread, step off along the walk line. The same number of steps as taken along the narrow end of the winder should bring the dividers close to but not beyond the other common tread.

Making a full-scale drawing

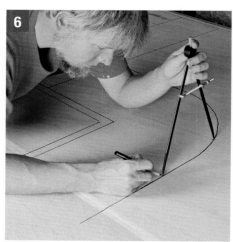

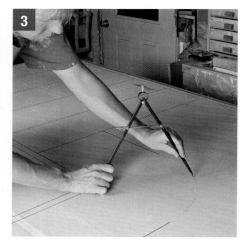

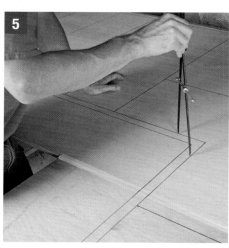

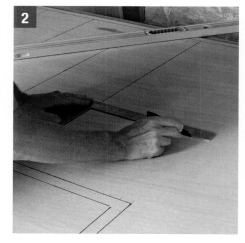

1 After drawing the perimeter of the stair on the plywood, pencil in the run of the upper treads. Don't forget to add in the thickness of the top riser.

2 Measure the walk line from the inside of the treads, and extend it far enough to catch all the winding treads.

3 At the corner, swing a 12-in. radius arc connecting the upper and lower walk lines.

4 Step off the steps with dividers or a beam compass set at 6 in. for the inside treads (you can use a larger dimension, but the resulting stairs will take up more space). Start at the lowest unit run mark on the upper straight stairs. For housed-stringer stairs, as these are, mark the inside of the inner stringer.

5 At the corner, simply swing the dividers to the perpendicular line.

6 Divide out the treads at the walk line. Set the dividers to the unit run of the stair, and step off along the walk line. Odds are that you'll have to experiment by opening the dividers a little to get the spacing right.

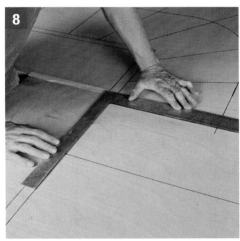

7 Connect the dots using a straightedge, extending the line all the way between the mortise depth lines on both stringers. These lines represent the front of the risers and the back of the treads.

8 If the last winder line is more than 90° off the inner stringer line, you've gone one too far. Erase that line, and draw one through the tread-depth point on the walk line that's square to the stringers.

If you end up beyond, you need to refigure the stair. Most likely, though, you just need to open the dividers a little and step it off a couple of times until the dividers land exactly on the common tread.

With two points of each tread marked, align a straightedge on each pair, and draw a line that extends between the stringers (photo 7). These lines represent the front of the riser above and the back of the tread below. The space between the lines is the unit run of the winder treads (photo 8). To lay out the actual treads later on, you'll draw lines parallel to these representing the nosing overhang.

Projecting elevations

Because of the varying geometry, laying out a winder's stringer without first drawing it is guesswork. Drawing out the stringers is necessary to determine if winding and common stringers can be cut from the same board, or if two boards will need to be joined for that section. The full-scale drawing you've done to this point already has all the unit runs on it, so all you need to do to draw the stringer in elevation is superimpose riser heights on the existing tread layout, producing what's called a projected elevation. This can be a little confusing, and it might help to visualize what you're about to do in this way. Imagine the plan-view drawing as being laid flat where the stair will go. Now imagine that the stringer is in place against the wall, and you rotate it down to lie flat on the plan view. If you did this, the stringer would lie flat on the plan view just as you're about to draw it.

Superimposing the riser heights starts at the line of the first common riser below the winding section. The idea is to measure in from the intersections of the riser lines and the outside edge of the plan-view stringer, marking the height of one additional unit rise for each riser line on the plan view. The first riser line represents a common riser, and it intersects the stringer at 90°. Measure in one unit rise from the edge of the stringer, and mark that point on the first riser line.

Seeing what the stairs will look like. Measuring square to the outside of the stringer where each riser intersects it, mark the riser heights on the drawing. Use these points and a drywall square to draw out the unit rise and run.

Drawing the wall-side stringers. Draw a line representing the top edge of the stringer 2 in. above the line between the top and bottom winder rise/run intersections. This space allows for the tread nosings. Do the same for any common treads.

Gauging the stringer depth. Using a framing square with its blade held to the line representing the top of the stringer, measure the depth required of the stringer. Measure to the bottom of the lowest riser, and don't forget to add a little depth so that the riser is fully housed in the stringer.

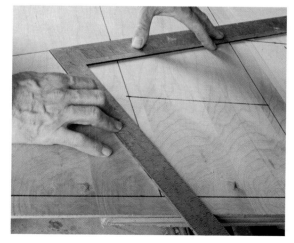

The next mark will be two unit rises in from the edge of the stringer, and the one after that, three. After the first riser line, the others are all winders and they intersect the stringer at increasing angles. Those points for those unit rises are not marked along the riser lines but rather square in from the edge of the plan view stringer (edge of plywood) where the riser lines intersect it. When you've marked one point for each riser, connect the points in a zigzag fashion by drawing in the unit rises and runs. The runs will be parallel with the plan-view stringer and the rises perpendicular to it. They're easily laid out using a drywall square held on the edge of the plywood, to draw the risers, and a framing square held to the drywall square to draw the runs.

Draw the stringer edges starting with the top edge. Mark one point 2 in. above the top riser and a similar point 2 in. above the bottom riser. Connect these dots using a long straightedge. This line isn't going to be parallel with the pitch line of the stair, but that doesn't matter. Measure down from this line to the bottom of the lowest hanging riser, and add about 1½ in. to that distance to determine the depth of the stringer stock. Mark a parallel line that distance below the first line to represent the bottom of the stringer.

Laying Out the Stringers

A single-piece stringer is easiest to make and strongest if you can lay it out to accommodate both winder and common steps. The starting point is deciding if the winders or the commons determine the slope of that stringer. On this stair, the fact that there's only one common tread means that both lower stringers can be one piece with the winders establishing their slope. Most of the common steps occur

Laying out a stringer

1 Measure the distances to the top of the stringer on the plywood. The pitch of the stair may not be parallel to the top of the stringer, although making it so simplifies layout.

2 Using the dimensions taken in step 1, pencil the first winding step and the riser above it onto the stringer stock, leaving room below for the commons.

3 Measuring the first winding tread's unit run from the upper riser drawn in step 2, hold the square congruent with the first winding tread's line and draw the riser below.

4 Align the square on the riser at the unit rise, and pencil on the common tread. Finish off laying out the common treads and risers in this manner, being exceptionally careful in aligning the square with the previously drawn unit rise or unit run lines.

5 Because of the shorter tread lengths on the inner stringers, it's a good idea to use two squares together. This approach avoids having to visually align the square on every previous line and helps to minimize errors.

Planning a joint in the stringers' geometry requires a joint, it's best to make it as unobtrusive as possible. In this case, continuing the line of a riser upward conceals all but the upper couple of inches of the joint.

Continue this procedure until the stringer is completely laid out. The common tread and riser can be laid out without referencing the top of the stringer by using a pair of squares, one aligned on the line of the closest winder (photos 4 and 5).

Making the Winder Treads

For the most part, stairbuilders ignore seasonal movement of the common treads. The angled grain of the stringer means this isn't truly a cross-grain connection, and problems are minimal. But winder treads are deep—approaching 2 ft. at the wall side. If they're made of solid lumber, you need to take seasonal expansion and contraction into consideration. This stair is a perfect example—the treads are hickory, which is one of the most active woods available. If it were built the same way as common stairs, it would rip itself apart at the first change of seasons. Different thinking is in order.

In this case, I used a combination of hickory treads and ¾-in. plywood subtreads. The subtreads, being plywood, don't move much. Glue and screws through the risers into the subtreads support their backs. The fronts of the subtreads rest on the risers below and are secured with glue and pocket screws. The hickory treads rest atop the plywood subtreads and are only pressure-fit in the stringer mortises. Screws and glue attach the backs of the hickory treads to the rear risers. The hickory treads are screwed to the subtreads, with the screws run through slots in the subtreads. This way, the hickory treads can shrink and expand seasonally, sliding over a rosin paper slip membrane that's between them and the subtreads. Cove molding below the tread's nosings hides the edge of the plywood subtread.

above the winders, so here, the commons determine the slope. The top inner stringer was made from one piece, as the narrow treads all fell within the width of the available stringer stock. The top wall stringer is joined because the steeper slope of the commons and the shallower slope of the wide ends of the winders couldn't be reconciled on a piece of stringer stock.

To lay out the winder stringers, start by selecting stock of sufficient depth and length. Lay a framing square on the projected elevation so that it's congruent with the first tread and riser, and note where the framing square crosses the line representing the top of the stringer (photo 1 on p. 119). Now, align the square on the stringer in the same way, and pencil on the tread and riser lines (photo 2). The riser height won't vary (photo 3), but the tread depth will.

Mind the grain direction. **Plywood is stronger along its length than across its breadth. Keep this in mind when laying out the subtreads, which should exactly match the unit runs from the full-scale drawing.**

Careful, but unguided, cuts are adequate for sub-treads. **Save the fine work for the parts of the stair that will show.**

Laying out the treads

The basic shape of the subtreads is already drawn out on the plan view. The only extra step that you need to take is to add another line to the end of each tread layout that represents the depth they'll penetrate the stringer mortises. With that information, lay out the subtreads on the plywood, keeping the treads more or less parallel with the strength axis of the plywood. The key to laying out these oddly shaped subtreads is to remember that the two edges that intersect the stringers will always be parallel. Keeping that in mind, you can transfer the corner points from the drawing to the sub-tread plywood stock with squares and measurements. Cut them out using a circular saw, and number them to avoid confusion.

Gluing up the treads

Given that it's unusual to find stock as wide as needed, nearly all winder treads need to be glued up. In this case, the hickory for the treads had been purchased as premade treads, cut to width and rounded over. Gluing two treads together back-to-back provided stock for two winder treads laid out so that their wide ends were at opposite ends of the glued-up stock. Subtreads serve as templates when

Subtreads are patterns for the show treads, which must be glued up. **To be sure the oddly shaped winder treads, which can be cut from the available stock, place two subtreads atop two lengths of tread. Rip a scrap to the width of the nosing overhang and hold it to each subtread when gauging its position. Mark the joint in the tread stock with a V for alignment during glue-up.**

preparing for the glue-up, and a scrap of stock the width of the tread overhang is added to the subtread width to ensure that the glued-up stock will indeed provide two treads. A couple of Vs drawn across the joint ensure that the stock can be positioned as laid out when the glue is spread. Use biscuits to properly align the tops of the treads.

After the glue sets up, I scrape off the excess and plane or sand the joint level. Then, use the subtreads and a scrap of wood the same width as the nosing overhang to lay out each tread to be cut. A circular saw is about the only tool for this job.

Slotting the subtreads

I use the time while the glue is setting up on the treads to rout slots in the subtreads for screws. The slots are about 1 in. long, and they're made with a ¼-in. straight router bit. Space them about 1 ft. apart across the tread's width and about every 6 in. across the tread's depth. The slots allow the tread to move independently of the subtreads, and a slip membrane of rosin paper between the two minimizes squeaks. I use wide-headed cabinet

Slots allow movement. Use a ¼-in. straight bit to slot the subtreads for screws. These slots allow treads to move relative to the plywood so they don't split.

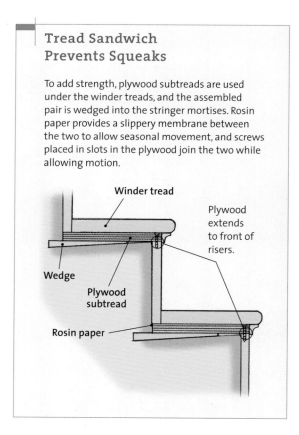

Preventing squeaks. Rosin paper, often used between hardwood flooring and the plywood subfloor, acts as a slip membrane to allow some movement between the subtread and the tread, while minimizing the potential for squeaks.

> ## Tread Sandwich Prevents Squeaks
>
> To add strength, plywood subtreads are used under the winder treads, and the assembled pair is wedged into the stringer mortises. Rosin paper provides a slippery membrane between the two to allow seasonal movement, and screws placed in slots in the plywood join the two while allowing motion.
>
>
>
> Winder tread
> Plywood extends to front of risers.
> Wedge
> Plywood subtread
> Rosin paper

Screws make the connection. Attach the subtread from below using 1½-in. pan-head cabinet screws. These screws should be roughly centered in the slots and set just enough to snug the subtread to the tread.

Molding covers a carpenter's sins. **The plywood subtread is fully supported by the riser below and will eventually be covered by cove molding.**

screws to connect the two. As you'll see later in this chapter, only the subtreads are glued to the stringers, so that the winder treads are free to move without cracking. The subtreads extend all the way to the face of the riser below and will later be covered with cove molding.

Making the Stringers

Winder stringers are made the same way as regular housed stringers, at least as far as the

Making a Long-Tread Jig

Like the standard mortising jig (see pp. 62–63), the winder tread jig is made from 1-in.-thick plywood. I make it approximately 30 in. long to accommodate maximum tread depths of about 2 ft. The cutout for the nosing is the thickness of the tread and overhangs 1¼ in., as on the other jig, but there's no riser. Instead, where the riser would meet the tread, the cutout is deepened by ¾ in. to allow for the plywood subtread. At the back end, the cutout is made to the combined depth of the tread, the ¾-in. plywood, and an additional 1 in. for the thick end of the wedge.

To lay out the angle of the bottom cut in the long tread jig, it's best to have a sample wedge that can be laid in place on the blank jig and scribed directly. Pencil this wedge onto the stock using a protractor, and screw an auxiliary plywood guide to the stock. Rip this wedge on a tablesaw, and use it to set up the auxiliary guide for subsequent cuts. After each wedge is cut, remove the stock from the guide and flip it end for end. Otherwise, the angles would add up, and later wedges would end up as fragile, cross-grained affairs.

Cut the first wedge to a measured size. **The butt of this wedge is 1 in., and it tapers to about ⅛ in. By screwing the wedge stock to a parallel-sided piece of plywood that acts as a guide, an accurate and safe cut is made.**

Replicating long wedges. **Use the first wedge to align the plywood guide for each succeeding wedge. Eyeball an allowance for the kerf. Once set up, the fence never moves, but the wedge stock is rescrewed to the plywood for each cut.**

Align the long jig's notch with the riser line. **The notch in the jig for the subtread lines up on the riser line drawn on the stringer.**

treads go. Treads, thickened with the addition of subtreads, are wedged into mortises cut into the stringers. Because of the angles involved though, mortising the stringers for risers just doesn't work well. The router cuts a perpendicular mortise, and the riser faces must be beveled to seat in the mortise. Each wedge would require special fitting, and a lot of time would be spent for a dubious reward.

Rout out the waste. As with any stair jig, rout in a clockwise direction to help maintain control. Several passes are needed to clean out the mortise's center.

Clean up the radius with a chisel. Inside corners of router-cut mortises always mirror the bit's radius. A couple of quick chisel cuts remedies the situation.

Rather than go through that rigmarole, I simply bevel the ends of the risers to fit to the stringer. They're attached with pocket screws from the back or through plugged pocket screws from the front. The choice depends on whether the riser encounters the stringer at an angle of more or less than 90°. If more than 90°, I use plugged pocket-screw holes from the front; if less than 90°, I simply pocket-screw from behind (as shown in the bottom right photo on the facing page).

To rout the tread mortises in the stringer, you'll need a special router jig, as well as the standard housed-stringer jig shown in chapter 4. Routing the mortises is pretty straight ahead. Lay the jig on the stringer so its top aligns with the tread layout, and the notch for the subtread lines up on the riser line. Then rout away in a clockwise direction. You'll need to chisel the front corner for the subtread square.

Building the Stair

Because their stringers ascend at different pitches, winders are astoundingly unwieldy as a unit. It's possible, but awkward, to build winders in one piece on a bench or horses. This approach wouldn't work on this particular stair, however. Because it was going into an older home, a typical application for winders, I had to work around obstacles that included an utter absence of anything else that was plumb, level, or square, and more vexing, a structural post in the corner where the stringers would meet.

Drawing it out

The first order of business was to make a sketch of the opening and the stair's geometry. That provided an approximate location for the bottom riser. Since neither the first-floor nor the second-floor landing were remotely level or flat, I measured the overall rise from the center of where the stairs would land on both floors. With an accurate overall rise in hand, I made the full-scale plywood drawing, which was used to dimension most of the stair parts.

Working from the bottom

I had to cut the bottom riser oversize and scribe it to fit the floor and be level. The heights of the bottom level cuts on the stringers derived from the scribed height, which differed by about ¾ in. from one side to the other of that riser.

I scribed the lower wall side stringer to fit between two timber-frame posts and the top riser to fit the unlevel floor. Other than these items, construction and installation was straightforward.

Who said stairbuilding was easy? Winders often go in old houses. This one wasn't square, level, or plumb, and a timber post interrupted the wall stringer. That stringer had to be marked and cut to fit before the lower stair was assembled. To do this, the stringer was held level in place and the cut line to fit it to the post scribed from behind.

Treads first. With the wide side on the floor for stability, the lower flight is wedged together. Note that there's no glue at the joint between the tread and the stringer. Only the subtread and wedge assemblies are glued to the stringers.

Risers second. Since these risers aren't wedged, take extra care in securing them to prevent squeaks. The holes for the screws are drilled using a pocket-screw bit, which creates a shoulder for the pan-head screws to bear on.

Assembling the upper stair

1 Reinforce joints in stringers with a gusset. Biscuits registered on the faces of the sections of the upper wall stringer ensure alignment, while a plywood gusset glued and screwed from the back ensures a tight and strong joint.

2 Slide the upper stringer onto the corner winding tread, shimming between the wall and the stringer. Shimming is needed to make sure that the tread is fully seated in the mortise.

3 Wedge the center winding tread home.

4 Install the first common tread to align the upper stringers. With all the angles going on here, it's easy for a winding stair's assembly to go awry. Clamping a common tread into place provides a control point for width, level, and alignment.

It's easiest to assemble winders piecemeal. The bottom section was put together on the floor next to the opening, then rolled into place, checked for level, and screwed to the wall. The inner stringer was propped level with a 2x4.

Assembling the upper stair

Assembling and leveling the lower stairs provides a platform to support the upper steps. They follow one piece at a time. This approach allows incremental tweaks to better fit the stairs to the opening. For example, you might not pick up that the well is ¾ in. out of square when you're first measuring. Build the stairs as a complete and square unit, and you've got to make a big fitting compromise somewhere. Build them in place, and you can make up ¾ in. by cheating each tread an unremarkable ¹⁄₁₆ in. Just be careful to keep the glue out of your hair when working.

Because of the length of the run involved and the huge difference in unit run between the four common treads at the top of the stair and the winder treads, the outside stringer for the upper flight had to be two pieces. That's not particularly difficult to lay out using the full-scale drawing and projecting elevations. Biscuits and a plywood gusset join the two stringers (photo 1 on the facing page).

With the corner winder tread already scribed to the post and installed in the bottom flight, I slipped the joined outside stringer into place (photo 2) and wedged it to the corner winding tread. A joined stringer is a pretty weak affair, so it must be securely fastened to a wall or otherwise supported. Note that glue is used only to join the wedge to the stringer and the subtread (photo 3). No glue is used on the tread itself, so it's free to move without splitting.

The inside stringer is placed and joined to its lower mate, and a common tread is slid into its mortises. A bar clamp draws the stringers tight, establishing the stair width (photo 4). Because of the angle of the winder, the nosings of the treads where they intersect the outside

Fitting the winder nosings. Because the winding treads intersect the stringers at angles, the overhang at the stringer is longer than the overhang measured perpendicular to the riser. You'd need a separate router jig for each tread to avoid this problem. Instead, slide the tread home, scribe it to fit the stringer, and notch it. Making the notch too long isn't a problem—just be sure to align the back of the tread with the riser line above.

Pinch sticks: the slickest trick going. Accurate inside measurements aren't easy, and they're made harder when they're angled. Take two scraps, and miter their ends to a point. Hold them between the surfaces you're fitting to and clamp.

Getting Around the Corner

Where two stringers meet at an inside corner, it's necessary to level out the upper one with an added block so that the trim can turn the corner.

Trim

Upper stringer

Level block

Lower stringer

stringer hit the back of the mortise and are open at the front. They must be held in place and marked so that the front of the nosing can be trimmed back. The length of the trim cut will match the distance that the untrimmed tread protrudes past the back riser, and the cut line itself can just be scribed along the stringer.

I used pinch sticks, a fancy phrase for two scraps of wood and a spring clamp, to measure the distance between the stringers. The angle is measured with a protractor or a bevel square, and the risers are cut to fit on a miter saw. Once installed, the risers are pocket-screwed to the treads above and to the stringers.

Screws join the risers and treads. **Alternate the screw patternat the risers' bottom between the back of the tread and subtread below and into the front of the sub-tread above. Use pocket screws between the riser and the upper tread.**

Build a supporting wall below the inner stringer, **as was done in chapter 5. Permanently screw the wall stringers from below into each stud, and clean up any glue drips.**

Winder Railings

Guardrails on winders don't differ much from guard-rails on stairs with a landing. Notching the newel is always a challenge, calling for patience and second- and third-guessing every cut (more on this in chapter 10). One difference between the two is that standard-length landing newels may be too short for use with winders. That's because of the additional rise gained with risers as opposed to a typical landing. If the newel terminates on the stringer, you should be able to buy one long enough with little trouble. If the newel extends to the floor, you may have to make or order a custom newel. If you're using a gooseneck fitting to drop down to the lower guardrail, be prepared for its drop to be longer than you're accustomed to.

If the inside stringer is notched and the balusters land on the treads, the baluster layout bears consideration at the time you're designing the stairs. As these treads are narrower than standard at this point—6 in. as opposed to at least 9 in. or 10 in., depending on your code—you might need to space the balusters on 3-in. centers. It all depends on the tread width at the baluster line. Don't forget that you can increase the tread depth of the common portions of the stair to improve this layout, assuming that you've got the run out at the bottom.

In the case of the stair shown in this chapter, the exist-ing structure severely limited the possible geometry of the stair. Happily, the owner wanted housed stringers. That meant the balusters sprang from the stringer and didn't have to bear any particular relationship to the treads.

Post-to-Post Railings

A lot of carpenters shy away from doing formal railing work. When you look at a balustrade and all of its angles and, yes, sometimes curves, it can be pretty intimidating. Railing work is considered its own trade by many, but like most stairbuilding, it's just an expansion of carpentry skills you probably already have. Not to oversimplify matters, but if you keep the newels plumb, the rails parallel to the stairs, and the balusters plumb and evenly spaced, you're more than halfway to a great railing job.

There are two main categories of stair railings: over the post and post to post. Over-the-post railings are never interrupted by a post. Instead, they pass over the top of the newels, using stock pieces of curved rail to change elevation and direction. Post-to-post railings run straight between two newels, which is where they change direction or angle. Now, there are hybrid configurations. It's not uncommon, for example, to start out a balustrade with a volute on a starting newel (over the post) but to have the railing butt into the intervening newels (post to post).

Stair Railings

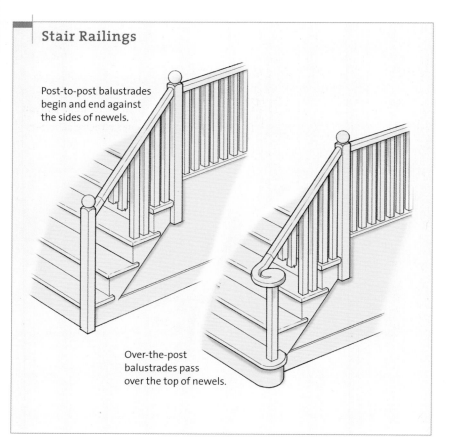

Post-to-post balustrades begin and end against the sides of newels.

Over-the-post balustrades pass over the top of newels.

Planning Out the Balustrade

The first step when installing any balustrade is setting the newels. Before you can do that, however, you need to determine railing height. Kind of a Catch-22, isn't it? When I'm doing an over-the-post system (see chapter 9), I noodle it out with a full-scale drawing, but that's mainly because of the curved components. Post-to-post systems aren't that complex.

Railing height is measured plumb up from an imaginary line that's tangent to (just touching) the tread nosings. That line is the pitch of the stair, and the pitch of the railing is parallel to it, about 36 in. higher. Fix these two lines in your head: one line at the nosings and one at the top of the railing.

Imagine a newel post at the front of the bottom step, and picture where these two lines intersect it. Now, slide that mental newel back

Safety First

The primary function of every railing is safety. Residential building codes distinguish between guardrails, which are intended to prevent people from falling over the edge, and handrails, which are intended to give stair-users guidance and a second chance should they stumble. On balconies 30 in. or more higher than the level below, a guardrail that's at least 36 in. high is required by the IRC. On stairs 30 in. or more higher than the level below, guardrails are required to be at least 34 in. above the tread nosings.

Handrails must be provided on one side of any stair of four or more risers, and they must be between 34 in. and 38 in. above the tread nosings. So, a guardrail can be a handrail. Balusters must be spaced so that a sphere of a 4 in. dia. can't pass between them.

The assembly must also be able to withstand a 200-lb. side load. To ensure the solidity of the balustrade, mount the newels and rails as securely as possible, and don't exceed the railing manufacturer's post-spacing recommendations. When I bolt or screw a newel into place, I use hardware that's long enough to penetrate the framing or the treads by about 3 in. When possible, I insert screws from two directions. After a newel is installed, it shouldn't move. Hit its top with your hand and you should hear and feel the vibration in the stair and the framing as a sort of deep thrumming sound.

on the step. If the newel remains the same height, by the time the newel is at the back of the step, the railing line is passing over its top. Next, imagine that same newel at the back of the top step. Slide it forward and what happens? If the newel doesn't get shorter, the rail line soon intersects the newel too low down. This mental exercise is the key to everything you'll ever need to know about setting stair railings. It applies not only to newels but also, using the line of the bottom of the rail instead of the top, to balusters.

Mental preparation

Here's another intimidating part of doing railing work: Before installation, just about every newel needs to be cut and notched to fit the stair. Some newels cost as little as $40, so trashing one is unlikely to break the bank. Some newels, however, cost 10 times that or more. Are your palms sweaty now?

Stair and railing parts are expensive. If you worry about that, you'll never unbox them. The best tip I know is to do the hardest parts first, while you're fresh and engaged. Then, double- and triple-check your layouts, and cut carefully while repeating this mantra: *It's only a piece of wood.* The alternative is to stand around half the day in fear of making a mistake and making no progress.

Start at the bottom

The simplest balustrades have only a top and bottom newel. I usually start a railing installation with the bottom newel. Since the height of a newel has to accommodate the railing height, and the newel height varies with the newel's relative location on the step, it makes sense that you must decide how far forward the newel will be on the step before cutting it

Post-to-post rails are the simplest balustrade configurations. **The rail goes directly between the newels with no curved volutes, easings, or goosenecks.**

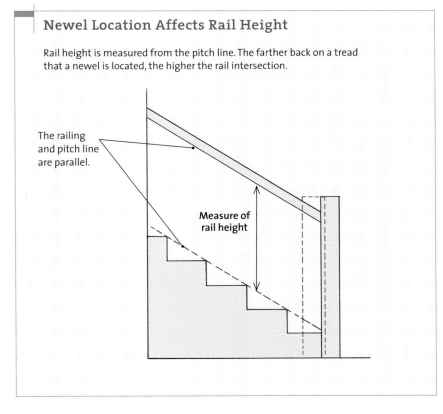

Newel Location Affects Rail Height

Rail height is measured from the pitch line. The farther back on a tread that a newel is located, the higher the rail intersection.

The railing and pitch line are parallel.

Measure of rail height

Don't Be a Slave to Symmetry

On a starting step, setting the newel symmetrically to the tread return and to the nosing overhang is nice. Sometimes, however, symmetry leads to an awkwardly placed baluster. It's more important to make the balusters look good than it is to keep the newel the same distance off the nosing as off the return. Landings are a different story, as the front-to-back location of a newel on the upper flight affects that newel's side-to-side location on the lower flight.

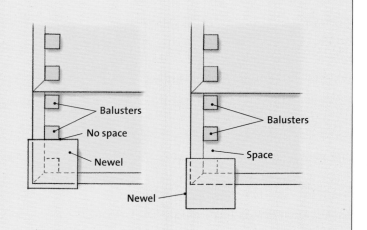

Balusters
No space
Newel

Balusters
Space
Newel

to height. Symmetry is also important, and it's particularly noticeable on bottom newels.

Symmetry plays into how far back the newel sits on the stair. On the first step, it's good to have the newel symmetrical to both the front and side overhangs. And since the balusters should center on the newel, how far the balusters are in from the side of the tread determines the newel's placement relative to the tread return. To maintain symmetry, I aim to set the newels so that their projection, if any, beyond the tread nosing equals that beyond the tread return. All of this may go some distance to explaining why railing work intimidates people. The balusters are the last part of a balustrade to be installed, yet they determine the location of the first part.

Be careful how far back you move a baluster to accommodate a newel. If you move it too far back, the rear baluster on that tread will end up too close to the nosing of the tread behind.

Locating the Starting Newel

Balusters are normally placed so that their outside faces align with the face of the finish stringer or skirtboard. Likewise, the front face of the first baluster on a tread is aligned with the face of the bottom riser. This spacing generally means that the bottom newel eliminates the first baluster.

Traditionally, and I think for good aesthetic reasons, newels either extend past or are flush with the nosings on both the front and the side. Newels that sit behind the nosing look like mistakes, unless the nosing and tread return wrap the newel, which is a lot of fussy work.

So, assuming a 3½-in.-wide newel, about the narrowest that I've ever installed, the center of the newel ends up at most 1¾ in. in from the edge of the tread return, whose point becomes the centerline for all of the balusters on that stair. That's a good location in another way, too. The ¾-in.-dia. holes in the treads for the dowels on the bottoms of the balusters just miss the tread return. The advantage of this is that the center spur on the bit will hit the stringer and not the cove molding. A spade bit hitting the cove molding can penetrate it or push the molding off the stair. And if the baluster hole ends up substantially in the tread return, then all that's holding the balusters in the event of someone falling against them is whatever is holding the return to the tread. It's far better to have most of the baluster anchored in the tread itself.

The newel shown here is the first newel on the stair, but it's also a landing newel. This stair has only one tread before a landing. As there was no railing below this newel, how it fell on the step below wasn't critical. But when I built and installed these stairs, the home-owners hadn't decided yet if they wanted a newel and a short rail on the first step. None is required by code, so this was an aesthetic deci-sion. Because the foyer where the stair begins is fairly small, the owners eventually decided not to crowd things with another newel. Nonetheless, I installed this newel no differ-ently than if there had been a lower newel. I think that alignment looks best, and it leaves open the option for the owners to add a lower newel should they change their minds.

Determining the height of the newel

Step one was to mark the newel's height. This newel needed to be that, plus the height of two additional risers to reach the floor. I bought a 6-ft. newel, marked its top for where the top of the railing needed to land, and set it on the floor next to the stair. Most newels have some sort of area for the rail to land on. Sometimes, it's a square block on a turned post. On this box newel, the area is delineated with mold-ing. It looks best if the rail centers on this area. To find the top of such a centered rail, first cut a piece of rail so that its end is plumb to the pitch of the stair, just as if you were fitting the rail to the newel. Measure the height of this cut, then measure the height of the rail block

on the newel. Half of the difference between them is the distance between the top of the block and the top of the rail.

I marked the edge of the first step where the back edge of the newel would go. Normally, I'd hold the newel next to its intended home to find its height. In this case, the lower flight was in the way, so I held the front of the newel on the mark instead of the back. However, the newel's face is in the same plane as its back will be after installation, so the information gar-nered becomes perfectly useful by rotating the newel 180°. With the newel held plumb and a level running down the tread nosings to estab-lish the pitch line, lightly pencil the pitch line of the stairs onto the newel at the lowest point.

The distance between the pitch line at its lowest point and the mark on the newel rep-resenting the top of the railing is the railing height. If it's lower than code minimum, the newel must either move forward (you gain roughly ¾ in. of railing height for every 1 in.

Mark the rail's top on the newel's upper face. Subtracting the height of the cut railing from the height of the newel block, then taking half of that dimension yields the distance between the top of the rail when centered and the top of the newel block.

Marking the pitch line of the stair on the newel provides a reference point to determine the newel height.

forward the newel moves) or you have to move the point where you want the rail to hit the railing. . . or you have to buy a longer newel.

Presumably none of those things happen, and the distance between the pitch line and where you intend the top of the railing to be is at least code minimum. If the distance is greater than you'd like, the difference between it and the height you favor is the amount to be cut from the bottom of the newel. Once the newel's height is known, go ahead and cut it. Having the newel at the correct height allows for some in-place marking, which I prefer to measuring when it's possible.

Marking and Notching the Treads

The next steps don't involve laying out the newel, as you might think. Before you can measure and mark the newel, you have to get the tread nosings out of the way. To me, these cuts are scarier than cutting the newel. Mess up a newel, and you buy a new newel. Messing up the stairs is more complicated. Fortunately, laying out and notching the treads is nowhere near as involved as laying out and notching a newel can be.

With a level clamped to the side of the newel, hold the newel in place and mark its edges on the treads. Because this newel intersects two treads and a landing, that sort of marking went on in several places. You really need a vision in your head of where you want the newel to end up. If you're not good at mental images, sketch it out from several angles. If you can draw it, you can build it.

Precisely marking cut lines on the treads is important, and marking the face of the tread by holding the newel in place gives extremely

Mark the upper tread for notching. **The same trick aligns the newel for marking on the upper tread, only this time the bottom of the level is registered against the face of the landing tread.**

Avoid measuring where you can. **A level clamped to the newel extends beyond its bottom, aligning the edge of the newel on the tread return below. In turn, this allows the newel itself to be used to mark the landing tread for notching.**

Getting past the roundover. **Accurately transferring the newel location from the face of a tread to its top calls for a pair of squares, one aligned with the mark on the tread face and the second held against the first.**

accurate results. But that gives an accurate mark on the front of the tread only, and you've got to transfer the line around the roundover to the top of the tread. Lining up a square on the mark by eye is one way, but it's more accurate to use two squares, one aligned vertically on the initial mark and the second butted to the first to guide the line.

A jigsaw makes quick work of the cross-grain cuts on the treads. I find a chisel to be faster to make the cuts along the grain. Cutting with the grain is slower by nature, and the fact that the cut comes flush with the riser or the stringer means that the blade rides along one or the other for the entire cut. Finally, somewhere near the front of the tread return is the screw that holds it in place. That'll wreak havoc with most sawblades, but cautious chisel work can proceed around the screw with no harm.

Once all the treads have been notched flush with the faces of the risers, you can start to lay out the newels.

Laying Out the Newel

An old joke asks, "How do you carve a stone elephant?" Get a piece of stone, and carve away everything that doesn't look like an elephant. That's about as encompassing a set of directions as I can give for laying out and notching a newel to fit a stair. Sure, there are a few techniques and a few things to watch out for, but in the end, you have to carve out the shape of the stair in the newel.

The notches in some newels are simple. For example, a newel in the center of a balcony just needs a block taken out of it to leave an L-shape that grabs the corner where the wall meets the floor. Others, like the one shown here, are more complex.

Sometimes, a newel seems so complex that you don't even know where to start. The example here might fit that description. In fact, I never comprehended this entire newel until it was cut out. I knew where it had to end up but couldn't visualize the whole thing. This is not

Crosscut the nosings using a jigsaw. **To avoid splintering, don't use an orbital setting but do use a new blade. Cut lines are visible on the nosings below the one being cut.**

A chisel makes quick work of the nosings. **Pay attention to the grain direction to avoid uncontrolled runout, and to the location of screws to avoid notching your chisel.**

With all of the nosings notched, **it's possible to begin to imagine notching the newel. The top tread here is notched to accommodate the newel wrapping around it.**

Laying out the newel

1 To facilitate measuring, draw a plumb line representing one edge of the newel on the wall and the stringer.

2 Measure from the benchmark. Measure the width of the bottom part of the newel between the level and the edge of the tread. The scale on a combination square is more accurate than a tape measure.

3 Mark the bottom notch on the newel. If the mating surface checks out as plumb, as the side of the stringer on the bottom step did, then setting a combination square to the correct dimension and running a line is fast and accurate.

4 Take two measurements on out-of-plumb surfaces. Here's a spot where reading the measurement would be difficult. Instead, place the square against the level and lock down the dimension. Sharp eyes not needed.

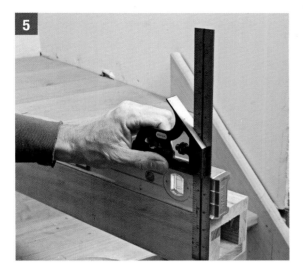

5 With the square positioned as when taking the measurement, mark one point of an out-of-plumb surface on the newel.

6 After the top and bottom points are plotted, connect the two with a straight line.

7 To find the pitch of the stair, measure with a protractor to a plumb line. The notch on the back of the newel must fit to the stringer.

8 Not all the layout lines on a newel are square. By reading the angle measurement on the protractor, you can be sure it hasn't changed.

an uncommon occurrence, and it makes a lot of people uncomfortable. Remember, it's only a piece of wood.

And really, you do ease into this, as all the lines are drawn and checked before any sawdust flies. I usually start at the bottom, laying out the cuts needed up to the top of the first step. One step's worth of cuts are easily comprehended. Only after laying out the horizontal cut that will rest atop the bottom tread did I start to figure out cuts that would get me up to the landing. With that horizontal cut drawn, I turned to the upper step.

Marking up the four sides

In more detail, here's how this particular newel went. The two main things the newel had to align with were the returns on the upper and the lower treads. Plumbing up from the lower tread's return determined where the newel would end on the upper tread's return (see the middle photo on p. 136). And plumbing down

Mark the height of the tread in place. The most accurate way to measure is often not to measure but to mark in place.

from the upper tread's return did the same on the lowest tread's return. To make measuring easier, I penciled the line of the newel's side on the wall. Tape measures have their place, but when it really counts, they aren't the most accurate tools. The rule on a high-quality combination square is far more accurate, and that's how I measure when possible (see the photos on pp. 138–139).

For a newel to end up plumb, its notches have to account for out-of-plumb stairs and framing. I check every mating surface for plumb before measuring. If a surface is plumb, then laying it out on the newel is a simple matter of setting a combination square to the correct dimension and using it to guide a pencil line. If a surface isn't plumb, then upper and lower measurements are taken and both are plotted onto the newel and connected with a line. When possible, the most accurate way to transfer a measurement is not to measure but rather to mark the work in place (see the photos at left).

Stairs have angles, and newels often have to fit them. I like using a protractor to measure and mark angles, as reading it gives me a number to use as a double-check when transferring the angle to the newel. A bevel square (also called a T-bevel) does the trick as well.

Repeat the same techniques as you wend your way up the newel. When it comes time to cut these lines you've drawn, it's easy to get confused and cut the wrong side of the line. To avoid this potential disaster, I mark the scrap with scribbles. I also take a few minutes to double-check the accuracy of all the layout lines. At this point, it's not only the cost of the newel at stake but also the value of all the time spent on the layout.

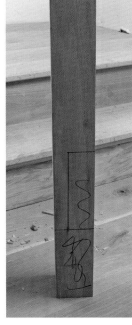

One line at a time results in complicated layout. **Before cutting, it's a good idea to double-check all of the marks on all four sides.**

Cutting and Fitting the Newel

Notching out the newel might be the most exciting moment in this entire process. It calls for good eyes, plus eye and ear protection, as well as a sharp blade. It also calls for concentration and understanding the purpose of each cut before making it. You have to reset the depth of cut for almost every cut. Make certain that the blade is square to the shoe, and this above all: Clamp the newel to something substantial, like the stair, before you fire up the saw. Cutting the newel is hard enough without it being a moving target.

Setting the depth of cut

When possible, I set the depth of cut by holding the saw next to a line that represents the bottom of a cut. Otherwise, I use a tape measure and err on the side of making the cut too shallow. Often, the notch crosses the entire width of a newel and so has to be cut from both sides. In this case, set the blade just slightly deeper than halfway. There's no point in going deeper, and doing so is hard on the blade and the saw.

Do You Need a Shooting Board?

I don't bother using a shooting board when cutting out the newel, as you might expect. When I started doing railings, I'd never heard of shooting boards and got confident enough that when they came onto my radar, they seemed cumbersome for this purpose. Essentially, it comes down to the fact that I've practiced enough to make the cuts quickly and quite accurately without one. But you should feel free to use a shooting board. They turn a circular saw into a precision tool.

Cutting is done by eye. It's not just cutting a straight line that matters, but stopping the cut at the right spot. It helps to set the blade no deeper than needed and to cut to the left of the line for visibility.

Plunge-cut slowly. Begin plunge cuts far enough back so that completing them would result in an overcut. Move the saw forward during the plunge so that completing it cuts just to the rear mark.

A utility knife makes for a clean cut. Before crosscutting, score the newel with a knife and a square to prevent the wood fibers from tearing out.

Making a plunge cut

Sometimes, a vertical cut has to stop at both the bottom and the top, requiring a plunge cut. Remember to pay attention to the back of the cut as you are plunge cutting. It's best to start the cut a little behind the center point, as if you planned to overcut the back line. As the blade penetrates deeper into the wood, move the saw gradually forward so that when the shoe is fully resting on the newel, the back of the blade just kisses the rear line. The reason

When cutting a newel, make as many cuts as possible with the saw positioned so that the cut line is to the outside of the blade and readily visible. Sometimes this isn't possible, but other times it's just a matter of making the cut from the other direction.

for this is safety. Moving a circular saw backward, particularly in hardwood, is asking for it to grab the material and kick back. When crosscutting, scoring the line first with a knife helps to prevent the wood from chipping.

Some circular saws, such as those made by Festool, are ideal for plunge cutting. They are designed so that you set the blade depth and then the blade and motor raise above the shoe until you're ready to cut. The big advantage, apart from the safety aspect of the blade always being above the sole until you're ready to cut, is that the sole of the saw remains on the workpiece, helping to align and stabilize the cut.

Clearing out the waste

After all the cuts are complete, it's time to clear out the waste. This is mostly chisel work (photo 1 on the facing page), some of it fairly crude whacking and splitting. When a notch doesn't go side to side but is enclosed, some serious chiseling is needed (photo 2). Because of the round sawblade, no stopped cut ever reaches the top of the notch except at the surface. The technique is to drive the chisel in at the corner of the horizontal cut, making a stop cut. Then you chisel the scrap down at an angle toward that horizontal cut (photo 3). When as much waste as possible is cleared out,

Clearing out the waste

1 Use a butt chisel to split out the waste. Don't try to take too much at once, as levering on the chisel can break it. The newel is resting on a short bench, and the author's weight keeps the newel from moving.

2 Start to remove the waste from notches with a stop cut. Heavy blows are needed, so be sure the surface the newel rests on is free of debris, like wood chips, that could dent it.

3 After the first stop cuts, remove waste with angled cuts. To minimize the length of the angled cuts, occasionally split out long chunks of waste.

4 Change chisels for the last little bits. Use a sharp paring chisel and hand pressure for final cleanup and for most adjustments that are found to be necessary while fitting the newel to the stair.

Looks complicated, doesn't it? **This would have been nearly impossible to visualize at once, but bit by bit, it came together.**

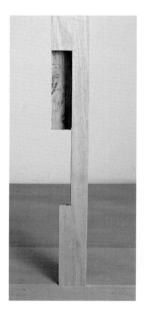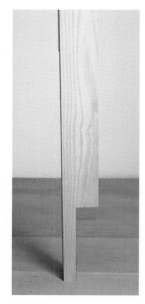

An oversize hole allows room for a hex driver and a plug. This ½-in.-dia. hole only goes in 1 in. or so and is followed up with a smaller bit for the pilot hole for the LedgerLok screw. A flat face-grain plug will fill the hole.

Impact driver sets screw. Setting the screw as high as possible in the newel provides better support against side loads.

you make more stop cuts. Every so often, make a splitting cut to make more working room and reduce the length of the slope you're chiseling.

A stout chisel is called for here. I use an inexpensive butt chisel ground to about a 35° angle before honing for this work. A bench or mortise chisel would also stand the pound-ing. I never use my paring chisel for this, as its slim body and acute grind wouldn't stand up. The paring chisel does make an appearance for trimming to the lines and for fine-tuning the fit (photo 4).

It's not unusual with a newel this complex to make a couple of trips back to the bench before the fit is perfect or as close to that as the day allows.

Attaching the newel

When the fit is good, I drill ½-in.-dia. holes with a new, sharp spade bit for the LedgerLok® screws used to attach the newel. (I use a minimum of four screws if the newel is only attaching to one step. In this case, the newel attached to three steps, and I used two screws per step.) These holes aren't pilot holes. Pilot holes come after the ½-in. hole, whose main purpose is to hide the head of the screw. Half-inch works well as the hex driver for the LedgerLoks just fits inside it, and plugs of that size are readily available or easily made with a plug cutter and a drill press. The holes are generally about 1 in. deep, sometimes shallower in a thin section of the newel. Depending on the substrate, I use either 3½-in. or 4½-in. screws. If there's a lot of meat behind the newel, I use the longer screws. If too long a screw is used though, the relatively short threaded section might run all the way out of the substrate, and the screw becomes useless and unremovable.

It's most important to firmly anchor the newel as high up as possible. Doing so reduces the length of the lever acting against the attachment. Getting a screw into the tread is ideal. Additional screws into the stringer and the underlying framing add to the stoutness. For looks, keep the screws equidistant from the newel's edges—about ¾ in. looks good.

Installing the landing treads

Depending on whether the hardwood-flooring crew got ahead of you, the last step in newel installation is the landing treads. On the newel shown here, the landing tread was in and I treated it like a tread on the stair. That wasn't the case with the upper newel, and I had to

LedgerLok Screws

Newels anchor the entire balustrade, so they must be firmly anchored. Lag bolts are one approach, but the holes needed to hide them are quite large, typically 1 in. to accommodate commonly available face-grain plugs. I prefer to use FastenMaster's® LedgerLok screws. They're as strong as or stronger than ½-in. lags, but the holes required to hide them are only ½ in. in diameter.

fit the landing tread to the newel (which is generally easier).

After setting the upper newel, I held the landing tread at the head of the stairs in place, marked it to length, and then installed it with 8d finish nails and construction adhesive. This landing tread extended past the back of the newel, so I mitered its rear corner to meet the short piece of landing tread that ran along the balcony.

Keeping squeaks at bay. **Bed the landing tread in construction adhesive, and secure it with 8d finish nails until the glue dries. The nails help, but the real attachment comes from the glue.**

Wide landing tread must be mitered to fit around the newel. **It's usually easiest if the flooring contractor tacks down the landing tread so that it can be removed and cut to fit around the newels.**

Two or Three Balusters?

The spacing between balusters is dictated by code and by aesthetics. Because I work in Connecticut, where 9-in. unit runs are legal and common, most of my stairs receive two balusters per tread. Two balusters on a 9-in. run spaces them at $4\frac{1}{2}$ in., and the balusters at their thinnest are $\frac{5}{8}$ in., providing a spacing of $3\frac{7}{8}$ in. In states that require a 10-in. unit run, you'll need either three balusters per tread or much thicker balusters.

Spacing Balcony Balusters

Let's say that your aim is an on-center baluster spacing of $4\frac{1}{2}$ in., and the space between the newel and the wall or the next newel on the balcony is $74\frac{3}{4}$ in. Divide $74\frac{3}{4}$ by $4\frac{1}{2}$, and you come up with 16.61. A whole number would have been sweet, but beggars can't be choosers. The next step is to round 16.61 up to 17, and divide $74\frac{3}{4}$ in. by that. The result is $4\frac{25}{64}$ in., or roughly $4\frac{3}{8}$ in. Because I rounded the fraction down by $\frac{1}{64}$ in., if the $4\frac{3}{8}$-in. spacing was used, the spacing between the last baluster and the last newel would be large by $\frac{17}{64}$ in., or nearly $\frac{1}{4}$ in. Not that anyone is likely to notice that, but splitting it and starting the layout $\frac{1}{8}$ in. farther away from the first newel makes it truly inconsequential.

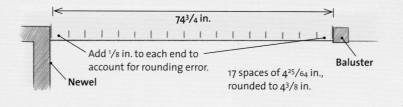

$74\frac{3}{4}$ in.

Add $\frac{1}{8}$ in. to each end to account for rounding error.

Newel

Baluster

17 spaces of $4\frac{25}{64}$ in., rounded to $4\frac{3}{8}$ in.

Balusters and Railings

Most of this chapter on railings has barely touched on them, per se. That's because compared with setting the newels, railings and balusters are a piece of cake. You've already put some thought into where the balusters on the treads are going in order to set the newels. Lay out the rest of the treads as you did the first and call that a fait accompli. The only baluster spacing that needs work now is on the level balcony (assuming your stair has one).

Spacing the balusters

How these balcony balusters lay out depends on a couple of things. Ideally, they'll be spaced so that the distance between the end balusters and the newels is the same as between all the other balusters, which should correspond to the spacing on the steps. In the words of my favorite Hollywood ogre, "Right. Like that's ever going to happen." In the real world, perfect baluster spacing is as rare as duck teeth. But you can come close if you've got room to work (see the sidebar at left).

Balusters will mount to the landing tread. **After figuring the baluster spacing, mark it on the landing tread. Mark the centers with a combination square set to center on the newel, and drill the baluster holes.**

One caveat here, and it's important. Turned newels can have parts that are substantially thinner than their square bases. If the baluster spacing next to these newels approaches the code maximum, and you start baluster spacing from the base and not the newel at its narrowest, the railing may well fail the 4-in.-sphere test. With turned newels, always measure from the newel's narrowest point.

Attaching the rail

Level rails fasten to newels and walls differently. If you're lucky, which I was in this house, there's solid framing where the rail hits the wall. It pays to take a look around if you're there to measure or install stairs at rough framing. Add blocking where you can see that it will be handy. Assuming that some other tradesman doesn't hack it out, you'll be glad for it. And so will the builder because railings can't fasten to drywall only.

In this case, the level rail landed solidly on a stud corner. (If that rail hadn't landed on solid framing, I would have added it from behind, opening up the drywall and installing 2x blocking between the studs.) I plumbed up from the baluster line and marked the top of the railing on the wall. As a marking tool, I use a thin section of railing with a pencil-size hole drilled in its center. This is a great tool, as it can be used to achieve consistent centers on the ends of railing and, by aligning it on a level line representing the bottom of the rail, on newels and walls as well.

Drill a ⅝-in. hole in the framing to accommodate a section of dowe. A corresponding hole drilled in the rail will slip onto the dowel. Next, drill a pilot hole in the newel and a matching pilot hole in the other end of the rail as well as ⅝-in. holes in the rail's bottom for

Lay out where the rail hits the wall. A center-drilled section of rail is used to mark for a dowel hole to be drilled in the wall. The rail section is also used to lay out centers for drilling the newel and the rail.

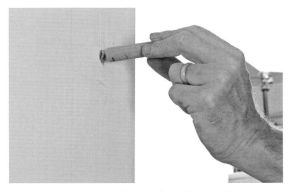

Hardwood dowel reinforces the rail-to-wall joint. Several inches of ⅝-in. dowel is inserted into the hole in the wall. A corresponding hole in the end of the rail completes the picture.

A bolt will affix the level rail to the newel. Drill a pilot hole partway through the newel. Complete the hole from the other side.

the balusters. The pilot hole is for a LedgerLok that's driven through the newel into the end of the level rail. Don't worry about plugging that hole because the handrail coming up the stair will cover it.

Fitting this rail was a bit of a challenge as it was so short there was no flex whatsoever. By slipping it onto the dowel and pulling the newel in the opposite direction, the rail went into place, but with a struggle. Toeing a couple of 8d finish nails through the railing into the framing by the dowel keeps it from spinning.

The dowel and LedgerLok combination is a great way to do a run of several balcony newels. One end of each rail is secured with a dowel and the other with a screw. By drilling the clearance hole for the screw the same ⅝ in. dia. as the dowel, one hole serves two

purposes. The clearance hole should go about halfway through the newel, and after the screw is set in one rail, the dowel for the next plugs the hole.

A combination of finesse and force. **Slip the rail over the dowel in the wall and pull back on the newel to horse the rail into place.**

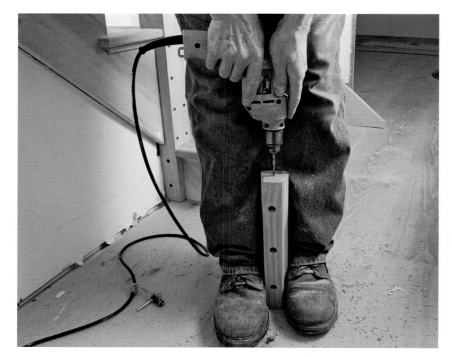

The rail requires a pilot hole. **Center-marked using the rail section, this level balcony rail is drilled to accept a LedgerLok screw set through the newel. Note the baluster holes in the rail's bottom.**

Impact driver sets the screw. **Driven into the pilot hole through a ½-in.-dia. countersink hole, a LedgerLok secures the balcony rail.**

Mortising a Newel for a Railing

Rather than use the dowel and LedgerLok combination described in the text, I sometimes attach level rails by mortising the newel to accept the rail. This approach does not work with pitched rails. I did this more often before the days of LedgerLoks, but it's still a viable approach.

Trace the profile of a piece of rail on the newel. Chuck a ¼-in. straight bit into a small router or a laminate trimmer and set it to take roughly a ⅜-in. cut. Start by routing around the perimeter of the tracing, staying about a bit's width away. This is key, as routing accurately to a line is nearly impossible if you're trying to do it with material contacting the other side of the bit. Such material contacting the spinning bit makes controlling the cut difficult. Get rid of it, and you'll be amazed how easy accuracy becomes. Focus on the edges now, taking little bites, sneaking up on them. Leave the pencil mark. Square up the corners with a chisel, and test-fit the rail.

Mortising a newel for the rail not only impresses the natives but also adds a fair amount of lateral support. I've seen other installers use this technique and a couple of toe nails, but I've always run in a couple of cabinet screws (*not* drywall screws) from the other side.

Trace the rail profile onto the newel.

Carefully rout to the line.

Square the corners with a paring chisel.

Test-fit the rail.

Marking and cutting the railings

Marking the railings for cutting is simple. Lay the railing in place on the nosings, mark it to length with a pencil, and cut on a miter saw. There aren't many tricks to this. I mark and cut the bottom of the rail first and lay it back on the stair with the bottom cut resting against

the bottom newel. That's when I mark the top cut. Initially, I cut the rail a little long and test-fit. Although the newels are set plumb, it's unlikely that they're perfect, so the distance between the top and bottom often varies a little. The rail gets trimmed as needed so that its top aligns with the top marks I've made on both newels.

Marking the rail for balusters

Before marking the rail for balusters, mark the baluster locations on the treads. Just continue the layout derived when you set the newels (see p. 134), and use a combination square to mark the distance in from the returns. To hold

A tape measure wouldn't work here. **After cutting the bottom of the rail, lay it on the stairs and mark the cut for the top. Plumb newels are important to this step, and cutting the rail a little long the first time is good insurance.**

Rail wants to roll. **When cutting a rail, be sure the flat bottom is fully seated against the saw's fence.**

Set the rail in place for baluster layout. **With a clamp supporting the bottom of the rail, use a level to plumb up from the baluster layout on the treads, and mark the side of the rail for the balusters.**

the rail for marking, place it on its marks, and fasten a clamp to support it on the bottom newel. Then, align the edge of a level with the baluster centers on the treads, plumb it, and lightly mark the edge of the rail with a pencil. Transfer these marks to the center of the rail's bottom using a combination square. Don't forget to erase or sand out the marks on the rail's edge. (In my experience, you can't rely on the painter to do this.)

Before taking the rail down, measure the baluster heights. The balusters should penetrate the rail by ¾ in. or so at the upper end of the hole. The next step is to cut the balusters to length. There's not much to this other than

Baluster tops must be fully housed in the rail. Measure the baluster height, allowing an additional ¾ in. to engage in the railing. With two balusters per tread, you'll have two measurements.

Buying Balusters

Balusters generally come in several heights, which vary an inch or so between makers. Common sizes are 34 in., 36 in., 39 in., and 41 in. The 34-in. balusters are used for the front balusters on a tread, as are the 36-in. ones. The 36-in. balusters are also used for most balconies, so you end up buying a lot of them. The 39-in. balusters are generally used for the rear balusters, and 41-in. balusters either below starting volutes (more on them in chapter 9) or for 42-in.-high balcony rails as are required in commercial work. The height designation of balusters includes ¾ in. for the pin at the baluster's bottom. The portion of the baluster from the tread up will always be ¾ in. shorter than its nominal dimension.

Why not buy all the tallest-size baluster you'll need and cut them down? First, because the balusters taper up to a point several inches below the top, from where they're turned to a uniform diameter, usually ⅝ in. That means you have several inches at the top of a baluster that can be trimmed off without increasing to a larger diameter. If you use, say, a 39-in. baluster at the front of a tread and trim it to height, you've likely cut off the ⅝-in. part and are into a taper. That means every baluster has to be checked for diameter before you drill its custom hole. Using balusters as they're intended, with graduated sizes, means that you can drill the rail for them using only a ⅝-in. bit. To add to the inconvenience, if you've cut into the tapered portion of the baluster, inserting it into its hole gets tough, unless you've drilled the hole oversize, which looks sloppy.

Second, the square bottom section of some balusters, as on the stair shown in this chapter, is taller on taller balusters. Start mixing balusters of varying heights, and the bottoms will run out of kilter. Finally, there's the matter of price. This isn't a big deal, as the tallest baluster might cost a dollar more than the shortest. Nonetheless, buy badly and you can leave the cost of several lattes in the scrap pile.

marking each one to length individually. Be sure to measure from the shoulder of the baluster that rests on the tread and not from the bottom of the baluster's dowels. This assures that you are measuring from the same point on the baluster and on the stair. It's also a good idea because the length of baluster dowels often isn't consistent.

With the balusters cut to length, verify their cut diameter. Take the railing down, flip it end for end, and lay it upside down on the stair so that the top of the railing is at the bottom. Chuck the appropriate spade bit into a drill (I always use a corded ⅜-in. drill for this operation). Make sure that the rail is seated on the tread nosings, steady it between your feet, and start drilling. Keep the drill as plumb as possible, but don't worry about it. Some variation here just doesn't matter.

The biggest concern is avoiding drilling through the top of the rail, while drilling a hole that's deep enough to allow the baluster to slip in ¾ in. farther than its final installed depth. (More on that extra ¾ in. in a couple of

The cut diameter of the baluster tops determines drill size. Balusters typically taper, so the diameter of the holes in the rail can't be determined with certainty until the balusters are cut.

paragraphs.) I check depth by eye, first holding the bit next to the rail at about the depth I want. I note how much of the bit's shoulder is buried at that point, and use that as my guide. Put a piece of tape on the bit as a guide, if you prefer. On steeper rails, you may have to begin drilling with the bit somewhat less than plumb. Center the point on the layout mark, start drilling, and move the drill up to plumb.

After the rail is drilled, chuck a ¾-in. bit in the drill and make the holes in the treads for the balusters. Once all the drilling is complete,

Drill the rail on the stair. **Set the rail upside down and backwards on the stair, and hold the drill as close to plumb as you can eye to create the baluster holes.**

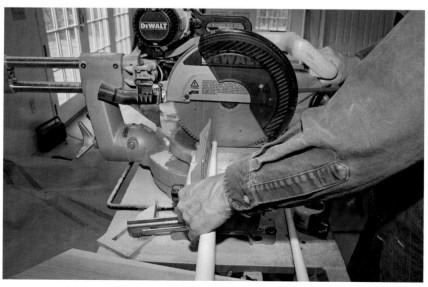

Use a new ³/₄-in. spade bit to make quick work of the baluster holes in the treads. Drill a little deeper than the baluster's dowels are long to leave room for excess glue.

Another Way of Attaching the Balusters

Another way to attach balusters is with double-ended lag screws. This technique works particularly well with custom-made square balusters. (Trying to add dowels to the bottom of a shopmade baluster is at best an exercise in frustration.) Slip a homemade drilling jig over the baluster to center the pilot holes in the balusters. The jig is just a plywood box that slips onto the end of the baluster. Its bottom is made from an inch or so of baluster stock that's center-drilled. Drill pilot holes in the treads, and use a special driver made of aluminum that's soft enough to thread onto the lags to set the screws. Then thread the balusters onto the lags by hand. The aluminum drivers are available from most railing suppliers, who also sell the double-ended lags in bulk. Both can be found in most big-box home centers as well.

Thread an aluminum driver onto the lags.

Set the lags in the treads.

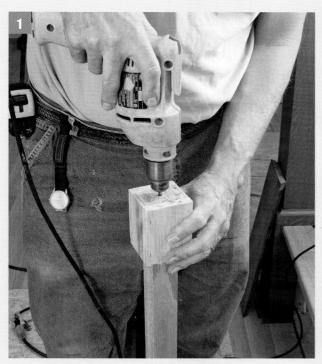

Use a plywood jig to center the drill to make pilot holes in the balusters.

Thread the baluster onto the lag by hand.

I run the drill at full speed, using the wind from its cooling fan to blow the chips out of the holes. This also serves to cool down the drill, which after drilling a number of holes in hardwood is pretty toasty.

Installing the rail and balusters

Once the balusters are ready to go and the rail is drilled to receive them, it's time to install the rail. This is a straightforward application of ½-in. clearance holes and 3½-in. LedgerLok screws. At the bottom, the screw enters from below, while at the top, it enters from above.

I start at the bottom and work up. With some yellow glue drooled into the holes in the treads, slip the balusters into the railing holes. When their bottom dowels clear the tread, position the balusters over the tread holes and pull them down. It helps to twist the balusters as you're working them into the rail, and if the holes are particularly tight, I chamfer the balusters before inserting them. Once the bottom dowels are seated, twist the baluster a bit to distribute the glue. Make sure the baluster seats fully in and is square to the tread.

I used to glue the tops of balusters but found little benefit and many troubles. Glue dripped onto stain-grade wood or didn't hold well on primed balusters. Sometimes, the heat from the friction of fitting a baluster would flash-set the glue before the baluster could be seated. Then, the baluster had to be cut off and the rail redrilled for another baluster, which I might or might not have had on hand.

Now, I just nail their tops. Balusters add little to no strength to the railing, and to be effective they just need not to fall out. Rattling disturbs people, and nails fix that. Drill a pilot hole for a 3d finish nail in the back of each

Bolt the rail to the newel. **Installed through a ½-in. clearance hole, a 3½-in. LedgerLok screw secures the rail to the newel.**

First up, then down. **Twist the balusters up into the rail far enough to swing them over the treads. Then drop the bottom dowel into the preglued hole in the tread.**

No glue in the top hole. Finish nails from behind secure the balusters to the rail. Try to nail downward slightly so as not to draw the baluster up and out of the hole in the tread.

You can't always nail where you want. Nailing through a pilot hole in the rail works well on level rails and when there's no clearance to work from behind.

Returned ends don't snag clothing. Wall rails are often needed to comply with the building code, and code-compliant wall rails have returned ends.

baluster after it's installed, then place and set the nail. Where there's no clearance for the drill or hammer, or where toenailing might pull the baluster up from the tread or floor, I drill and nail from the side of the rail.

Installing Wall Rails

In many cases, the railing on the outside of the stairs is all that's needed to serve as both a guardrail and a handrail. Other times, a handrail is required by code along the wall. All handrails must terminate safely, and mostly that's interpreted to mean that they return to the wall at the top and bottom. To achieve this end, the rail is mitered top and bottom so that it's long enough to extend at least beyond the upper and lower tread nosings of the flight of stairs.

Code requires at least 1½ in. of clearance between the rail and the wall, so I cut the returns to 1¾ in. to leave some room for trimming. So they don't loosen with time, drill pilot holes and fix the returns to the rail with five-minute epoxy and 8d finish nails.

Attaching the wall rail

Attaching wall rails is a little harder than it might seem on first read. The trick is locating the brackets at the right height on studs. By laying the rail atop the wall stringer (photo 1 on p. 157) and measuring to its top from a tread nosing, you find how much needs to

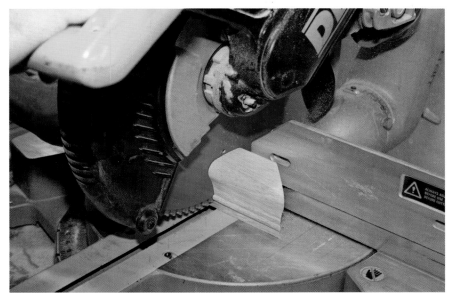

Work with longer stock to keep your fingers away from the saw. It's safest to cut a return by first mitering a long rail section, then cutting the return from it.

Predrill for finish nails. With the rail and the return seated on a flat surface, drill pilot holes. Go slow and hold the parts with your fingers so that the drill doesn't push them apart.

Work several nails at once to prevent the pieces from moving out of alignment. Start the nails into their pilot holes and hammer each one partially home before setting any of them. Five-minute epoxy ensures a long-lasting joint between the returns and the rail.

be deducted from the desired top-of-the-rail height to locate the top of the bracket from the stringer. By deducting from that the distance between the bracket's top and its bottom screw hole, you get the distance between the top of the wall stringer and the bottom screw hole in the bracket. Mark this measurement on the side of a 4-ft. level, and use the level to quickly and accurately mark the bracket locations on the wall (photo 2). Because your deductions accounted for the stringer height, the level can rest anywhere on the stringer where there's a stud. I find studs by tapping the wall with a knuckle or getting my eye close to the wall and looking parallel to it. Drywall fasteners are usually quite visible this way. If I was smart when I installed the stairs, I marked the studs' locations on the stringer.

Once you've located the stud, mark the bracket location, drill a pilot hole, and screw the bracket to the wall (photo 3). With both brackets mounted, hold the railing in place so that it rests fully on them. Check the returns for a good fit to the wall, and scribe them if need be (photo 4). Trimming to the scribe line on a miter saw nearly always requires a helper or a stand to support the end of the rail because there's so little of it on the saw table. With the returns pushed to fit nicely to the wall, affix the railing to the brackets (photo 5), and the day is done.

Attaching a wall rail

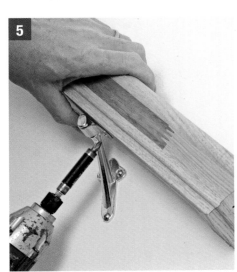

1 Figure the height of the rail brackets. Measure the combined height of the rail and the stringer from the nosing. Deduct this number from the desired railing height to find the bracket height as measured from atop the stringer.

2 Pencil the bracket height, less the distance from the bracket's top to the bottom screw hole, on the level. Align that with a stud, and mark the bracket's bottom screw location on the wall.

3 Affix the wall bracket. You may think there's a stud there, but until the drill brings up wood chips, you don't know. Bracket screws can break, and pilot holes help to avoid that.

4 Walls are rarely perfect, and scribe-fitting the return is typical. When scribing, be sure to keep the flat of the rail bottom seated on the bracket. If you don't, the mounting screws will pull the rail flat anyway, and your scribe won't be so noteworthy anymore.

5 Seat the rail on the bracket, and drill pilot holes for the bracket screws. Keep the rail pushed tight to the wall and fasten to the bracket.

Over-the-Post Railings

Back in the 1980s, the first railing jobs I tackled were pretty simple post-to-post balustrades, and I avoided over-the-post systems for at least the first year I was doing stair and rail work. Then a builder I was doing a lot of other work for twisted my arm into trying an over-the-post job. As is often the case when traveling in new territory, I imagined more troubles than reality threw at me. The knowledge I'd picked up doing post-to-post rails combined with the instructions that came with the stair parts kept me out of the really hot water.

Start with a Full-Scale Drawing

It took an embarrassingly long time for me to figure out the easy way to do over-the-post railings, which is to plan them out full scale. Prior to that, I used a lot of educated guesses and straight-edges run between newels to represent railings to help noodle out a railing. That's the curse of the impatient: You spend a lot of time just doing it, then you take it apart and redo it until it's right. Now I know that investing some time up front in planning pays big dividends in time saved doing the installation.

Anatomy of an Over-the-Top Balustrade

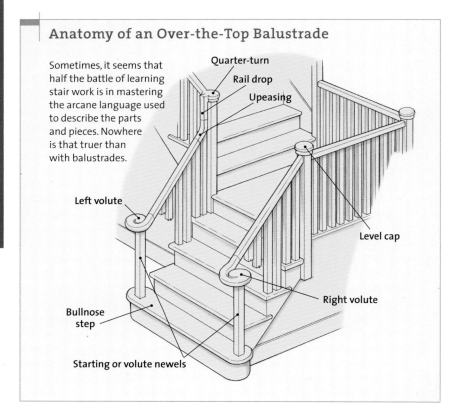

Sometimes, it seems that half the battle of learning stair work is in mastering the arcane language used to describe the parts and pieces. Nowhere is that truer than with balustrades.

Quarter-turn

Rail drop

Upeasing

Left volute

Level cap

Bullnose step

Right volute

Starting or volute newels

That said, I don't do a full-scale drawing of the entire stair, just those parts around the newels where all the fittings go. I lay them out in detail on a sheet of inexpensive ¼-in. plywood or Masonite. Using a 4-ft. drywall square in concert with a framing square facilitates layout, in much the same way that a drafting table with a parallel bar facilitates mechanical drawing.

The drawing of the stair is a side elevation that includes the tread nosings. All stairs, even those that turn at landings like the one shown, are simply drawn in two dimensions. In that case, the landing shows up as two unit rises without a unit run, as if the 90° inside corner

If you can draw it, you can build it. A 4x8 sheet of inexpensive plywood makes a perfect pallet for laying out the balustrade. There's no need to draw the entire stair, just the areas with newels.

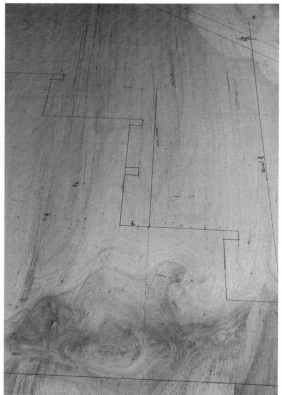

Flatten the turns. Landings are drawn as if they'd been opened up and laid flat. Two risers and no tread indicate the landing's position.

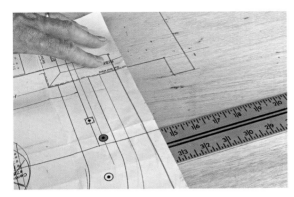

Begin with the starting newel. Lay out the location of the starting newel—the one that will support the volute—using the volute's template and a drywall square. Note the riser line on the template coinciding with the riser on the drawing.

of the landing had been opened up flat to 180°. This presentation gives an accurate picture of the length of the landing newel.

After drawing the stairs themselves, measure up square from the tread nosings to the desired railing height at the top and bottom of each flight. Connect these marks with lines, and draw a second, parallel line representing the bottom of the rail below the first line. On stairs with landings, draw the railing twice: once parallel with the stair up to the landing and once parallel with the stair parallel above the landing. The two railing lines will be parallel and separated perpendicular to the treads by one unit rise. Using the template provided with the starting volute, locate the center of the starting newel on the tread and extend a line upward from that point to the level of the railing.

Cutting a Volute

Over-the-post balustrades use a variety of fittings—curved parts that gracefully turn corners. Because stock stair parts like volutes and upeasings (transitional fittings that sweep upward) have to accommodate a range of stair pitches that vary with the rise and run, they're made long and need to be cut to length. For a fitting to join a piece of straight railing correctly and smoothly, the joint has to occur at a point where the line of the railing is tangent to the fitting, and both the railing and the fitting must be cut square to this line. Cutting the railing square is a no-brainer—you set a miter saw to 90° in both axes and cut the railing. Marking and cutting the fitting is the trick, and knowing how to do this removes about half the mystery from over-the-post railings.

Marking the volute

The key to this puzzle is a "pitch block," which is simply a piece of wood cut into a right triangle with one leg representing the height of the unit rise and the other leg the length of the unit run. The pitch block will find both the tangent point and the cut angle on the fitting. To mark a volute, clamp it on a flat surface and slide the pitch block, with the run side down, under the volute's upeasing. The contact point between the two is the tangent point. In a pure geometric world, this point should be distinct—just the merest contact. In the real world, the contact point often looks more like the contact quarter inch. I get down to where

Don't discount the instructions that come with the stair parts. There's a fair amount of minutia in stairbuilding, and it's easy to forget something like how far back from the edge of a rail fitting to drill the bolt access hole. That information is on every set of fitting instructions I've seen.

A pitch block can be a scrap from notching the stringer. **Where the block touches the volute is where that stair pitch is tangent to the curve of that easing.**

Flip the pitch block onto its rise **to mark the cut line through the tangent point. This cut line is square to the pitch of the stair and will allow a square-cut railing to joint to the volute.**

Pitch Blocks

A pitch block **is a piece of wood cut into a right triangle (you can use the scrap from notching the stringer). One leg is the height of the unit rise, and the other leg is the length of the unit run. The length of the hypotenuse is what it is—what matters is that it's nice and straight. The pitch block shown in the photos at above is a scrap of 3/4-in. plywood, and it works nicely because it will stand up on its own. In a pinch on job sites, I've often made pitch blocks by cutting up a cardboard box.**

I can look squarely at the pitch block and the fitting, note the ends of that quarter inch, and mark its center as nearly as possible by eye with a pencil on the fitting.

With the tangent point marked, stand the pitch block up on its rise, align the hypotenuse with the tangent point, and pencil the cut line onto the volute.

Cutting the volute

Now comes the part that some folks might find tricky—cutting the volute. It's possible to place the volute on edge on the miter-saw table, with the bottom tight to the fence, align the cut line

Support volutes with a cradle. **A few minutes spent tracing the curve of the volute's upeasing onto a block and bandsawing out the shape pays dividends in both safety and accuracy when it comes time to cut.**

with the blade, and with a firm, confident grip on the volute, slowly make the cut. I've made many cuts this way, but my fingers are closer to the blade than seems strictly safe. Another stairbuilder I know, Stan Foster, has cut a cradle to the same radius as the easing, which he uses to support volutes as they're being cut. I use a version of Stan's cradle now.

Tenoning the Newel

After cutting the volute, take it back to the plywood drawing and clamp it to a homemade right-angle bracket that lines up on the volute's center hole and on the centerline of the newel.

Adjust this assembly up or down until the cut end of the volute aligns with the railing lines on the drawing. When the alignment is right, mark the plywood where the volute's bottom intersects the newel centerline. The distance between this point and the bottom of the first riser is the newel's overall height. The unit rise is the height of the tenon that fits into the bullnose step; mark this measurement all the way around the newel's base.

While it would certainly be possible, and a bit less work, to make the mortise in the bullnose step the same size as the newel, there are advantages to cutting a shoulder and tenoning

The bottom of the volute locates the top of the newel. **Mark the newel's bottom with its height and the depth of the first step. Those are the top and bottom of the tenon to be cut.**

Center the volute on the starting newel. **With the volute clamped to a right-angle brace, lay it on the drawing so the centerline of the newel intersects the hole in the volute and the right-angle brace is right on the newel's centerline.**

The volute's cut should line up on the rail. **With that so and the center hole over the newel's centerline, mark the volute's bottom on the drawing.**

Tenoning the starting newel

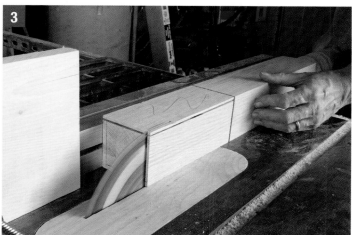

1 Set the depth of cut on the sliding miter saw for the tenon shoulders.

2 Cut the shoulders to the depth of the tenon. A stop block clamped to the saw behind the blade ensures that the shoulder cuts match all the way around the newel.

3 Cut the tenon's cheeks on a tablesaw. A block clamped to the fence prevents overcutting. When the work hits the block, shut the saw down. When the blade is still, pull the newel back, rotate 90°, and begin another cut.

4 Bandsaw through what the tablesaw leaves to finish the cheek cuts. No fence is needed as the tablesaw kerf guides the bandsaw.

5 Test-fit the newel in the bullnose step. It's likely some additional stock removal with a chisel, rasp, or plane will be needed for a perfect fit.

the bottom of the newel to a smaller size. In the first place, it's difficult to make a perfectly clean mortise cut in the top of the tread. The shoulder hides a few sins here. The shoulder also provides a positive stop so you know that the newel is properly seated.

To make the shoulder cuts, I limit the depth of cut on my sliding miter saw to leave the tenon the desired width. A stop block to one side of the blade ensures that the cuts made on all four sides of the newel line up perfectly. To trim the sides of the tenon, set the tablesaw to the proper cut and clamp a stop block to the fence so that the rip cut doesn't pass the shoulder cut. When the newel hits the stop block, do not pull it backward with the saw's blade still spinning. That would invite kickback, where the sawblade throws the workpiece back toward the operator with incredible force. Rather, shut the saw down with the work still engaged, and wait for the blade to fully stop before moving. Finish the cuts begun on the tablesaw with a bandsaw (a handsaw, jigsaw, or chisel would also do the trick).

Some cleanup of the cuts with a plane, rasp, or what have you may be needed. The newel should fit tightly into its mortise. When test-fitting, high spots on the tenon become slightly burnished, and their higher gloss is easily visible with a flat light source. Take these spots down, and try again until the newel dry-fits the bullnose.

Goosenecks and Quarter-Turns

A gooseneck is a rail fitting that accommodates the extra rise imposed on a balustrade by a landing. In essence, when a stair turns at a landing, its inside corner ascends two unit rises without an intervening tread, or unit run. Because there's no run at this point, the pitch of the railing changes to vertical. Another way

of looking at this can be seen in the full-scale drawing of this chapter's balustrade. The upper and lower railings, if extended past the newel, would have the same pitch, but they'd miss intersecting each other by one unit rise.

This change in the railing's pitch can be accommodated in several ways. If you're using a box newel (see pp. 182–187), or a turned landing newel with a large upper block, both the upper and lower railings can intersect the newel. This is a simple way to handle landings. Frequently though, even post-to-post balustrades employ a gooseneck on the downhill side of the newel. Goosenecks in this iteration spring straight out from the face of the newel, turn down, and meet the railing coming up from below with a curved piece called an upeasing.

You can buy a 90° downturn, but they're pretty simple to make, too. Miter-cut two pieces of rail, and slot them for #20 biscuits. The short side springs horizontally from the newel. It's usually trimmed to about 2 in. long, but make it 3 in. or more long to start. The

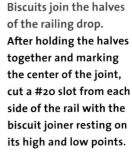

Goosenecks make up the additional rise at landings. A railing drop and an upeasing comprise a gooseneck fitting, which is frequently found on all sorts of balustrades.

Biscuits join the halves of the railing drop. After holding the halves together and marking the center of the joint, cut a #20 slot from each side of the rail with the biscuit joiner resting on its high and low points.

Epoxy and #20 biscuits make a strong joint. Since assembly goes pretty quickly from this point, 5- or 10-minute epoxy provides plenty of open time.

Miter clamps secure the joint until the glue sets up. Leaving only a couple of finish nail-size holes to fill, Clam Clamps pull corners together outstandingly well.

long, vertical part—the drop—should be 10 in. or so. Liberally coat the biscuits and the faces of the rail with epoxy. Assemble the pieces, and hold them together with a miter clamp.

To mark the easing on the bottom of the gooseneck for cutting, use a pitch block as with a volute, with one important difference. Because a gooseneck and a volute are oriented 90° differently from each other, you need to mark the tangent point on a gooseneck with the pitch block standing on its rise. Mark the cut line with the pitch block resting on its run, which is the opposite procedure from laying out a volute for cutting.

Cutting and joining a quarter-turn and a drop

True post-to-post balustrades have a newel with a doweled top and what's called a quarter-turn fitting to round the corner at the landing. As with the box newel shown in chapter 8, landing newels are taller than other newels, as they extend down at least to the top tread of the lower flight of stairs.

The quarter-turn must be cut and fitted to a vertical railing drop, just as a gooseneck fitting is done, and its upper easing must be cut and attached in order to plot out the height of the landing newel. Miter cutting the quarter-turn is a bit more challenging than miter cutting straight rail. First, you must decide how close to the newel you want the rail drop to be. I like it to be close—about 1 in. away from a line extended upward from the newel's square bottom. Keeping the drop, and by extension the lower easing, close to the newel ensures that the upper baluster will be tall enough to reach the easing.

If, for example, the square part of the newel measures 3½ in. wide, and you want the vertical drop to be 1 in. out from that, the bottom of the quarter-turn would be marked at 2¾ in. (half the newel width is 1¾ in., plus 1 in. for clearance) from the center of the quarter-turn's dowel hole. From that point, extend a 45° line upward to the top of the fitting.

Here's where it gets a little tricky. Because of the quarter-turn's bulbous shape, it can't simply be tossed on the saw and mitered. I set the saw's bevel to 45° and clamp a sacrificial spacer block to the fence to support the straight part of the fitting. This is too small a piece to hand-hold for the cut, so I use my miter saw's hold-down clamp. The trouble with this is that the hold-down on many miter saws doesn't have

Cutting a quarter-turn

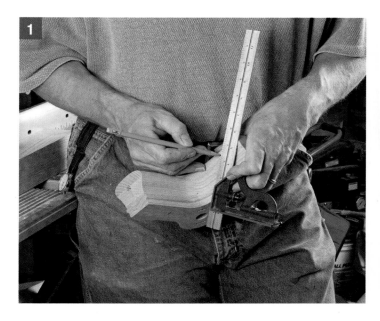

1 A quarter-turn cap rounds the corner at the landing. Find the back of the quarter-turn's miter, where it will join to a rail drop, by measuring the desired distance from the dowel hole. Mark the cut on the fitting using a combination square.

2 Quarter-turns don't sit well on miter saws. With the straight edge of the quarter-turn blocked away from the fence, line up the cut on the blade.

3 This isn't a cut to be handheld, so use a hold-down to clamp the quarter-turn to the saw for safety and accuracy. Note the block to the left and the bridging piece of wood that actually bears on the quarter-turn.

A thin, center-drilled section of rail **cut on a miter locates the centers of dowel holes in the quarter-turn and in the vertical drop.**

Drilling square to an oddly shaped workpiece. **Watch for even contact when the bit hits the fitting to ensure a dowel hole drilled square. Adjust the drill angle as needed, then press it in.**

Watch that this joint doesn't twist. **With a ⁵⁄₈-in. dowel and epoxy between the pieces, a miter clamp secures the joint. Until the epoxy sets, the dowel acts like a pivot, so you have to pay attention to how the parts seat when tightening the clamp.**

enough reach to clamp something this close to the sawblade. Yankee engineering once again saves the day. By placing a block of approximately the same thickness as the fitting on the end of the saw table and bridging from it to the fitting with a stout piece of stock, the hold-down can be employed.

The quarter-turn's bulb prevents the use of a biscuit joiner. In this case, I use a ⁵⁄₈-in. dowel instead to align and reinforce the joint. A thin section of rail cut at 45° and center drilled serves to mark both the quarter-turn and the rail drop for the dowel hole. I drill both the fit-

ting and the rail by eye for the dowel. There's a trick to drilling holes square to the material using a spade bit. Hold the drill as close to square as you can, and watch how the bit's outer spurs first hit the wood. If the drill is square to the material, an even amount will be removed all around. Cut a dowel, test-assemble the pieces, then flood the dowel and the rail surfaces with epoxy. As with the previous case, use a miter clamp to draw the work together.

Joining Easings to Rails

To transistion to the rail on the uphill side of a quarter turn, an upeasing is used. Before attaching the upeasing to the quarter-turn, you have to decide how close to the newel the curve begins. The farther back the joint is, the higher the newel will have to rise to keep the rail above code minimum. If the joint is too close, you'll have a hard time installing the

Rail bolts often come with a star-shaped nut that's meant to be tightened by tapping on it with a hammer and nail set, rather than a hex nut that tightens with a box wrench. These star nuts make great fishing sinkers, and they skip across a pond just like a flat rock. That's about the extent of their usefulness, though. If your rail bolts come with star nuts, replace them with ⁵⁄₁₆-in. coarse-thread hex nuts.

bolt that holds it together. When all else fails, I usually let symmetry make my carpentry choices, so I cut the quarter-turn's uphill end back the same distance (2¾ in.) from the dowel hole as its downhill end.

The classic way of joining pieces of rail to fittings such as volutes and easings is with a rail bolt. It's a ⁵⁄₁₆-in.-dia. hybrid that's a lag on one end and a machine thread on the other. The lag end is driven into a pilot hole in the curved fitting, while the machine bolt end enters a clearance hole drilled into the straight fitting or rail. An intersecting 1-in.-dia. hole drilled into the bottom of the rail or straight fitting provides access to fit and tighten a washer and nut. This hole is filled with a plug that comes with the rail bolt.

Drilling the ends

Center the 1-in.-dia. hole needed in the rail or quarter-turn on its bottom, 1⅜ in. from the face (see photo 1 on p. 170). It has to be 1½ in. deep to provide clearance for the nut and wrench. Be careful not to overdrill this hole: It's easy to come out the top.

Cut a thin section of railing to use as a center marker to mark holes on both ends to be joined (photo 2). Drill the center of the marker ¹⁵⁄₁₆ in. up from the bottom with a hole that's

big enough to accommodate a pencil point (photo 3). Now, it's almost unheard of to find rail and fittings that are cut to exactly the same profile, so to use this center marker, just align it so that it fits as well as possible all around. The differences in the profiles will be faired out later. Similarly, it's nearly impossible to perfectly center the hole in the center marker. Minimize the effect of being off a little here by holding opposite sides of the center marker to the pieces being joined. That way, the hole's position is exactly lined up on the opposing pieces. If you align the center marker with the same side against each piece, any deviation from center in where you drilled the hole is doubled.

After marking the faces to be joined, start drilling with a smaller bit, say, ⅛ in. That helps a lot with accuracy (photo 4). I find that no matter the diameter of the twist bit I'm using, it's easy to be off my mark by half the bit diameter. Well, half the diameter of a ⅛-in. bit is pretty inconsequential, and the larger bits just follow the path of the smaller starter bit. The hole in the easing is ¼ in. in diameter, and since the bit's chucked in the drill already, I drill a ¼-in. hole in the rail or the quarter-turn as well (photo 5). That's only a precursor to the ⅜-in.-dia. hole needed in the rail or quarter-turn, though.

Driving the rail bolt

A pair of lockjaw pliers works to drive the rail bolt into the easing; use some wax on the threads to ease this operation (photo 6). If you forget the lockjaws back in the shop, an option is to snug two nuts together on the machine thread and use the outer one and a wrench to drive the lag. A third option, which is pretty slick, is to use the specialty wrench sold by

Joining fittings with rail bolts

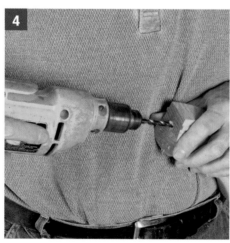

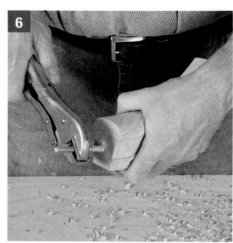

1 Drill a 1-in.-dia. hole in the bottom of the quarter-turn (or the railing) for access to the rail bolt.

2 Drill the center of a railing section, 15/16 in. up from the bottom, to use for laying out the bolt holes.

3 After marking the ends of the parts, drill 1/8-in. pilot holes to ensure accuracy.

4 Hold the drill as square as possible to the end of the easing while drilling the hole for the lag end of the bolt.

5 Drill the clearance hole for the bolt with a 3/8-in. bit.

6 Drive the bolt into the easing with locking pliers set far enough down the machine thread so as not to mar them where the nut will go.

7 Alternatively, drive the bolt into the easing with a specialty wrench from L. J. Smith, which has a nut welded to it for driving rail bolts.

8 A hemicylindrical washer gives the nut solid bearing. Matching the radius of the 1-in. hole, the washer doesn't push the parts out of alignment.

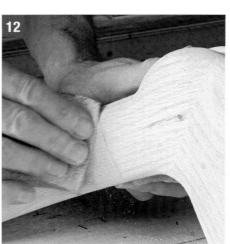

9 If using regular flat washers, chisel a flat around the bolt-clearance hole to provide bearing.

10 After coating the ends of the parts with epoxy, use a wrench to tighten the nut and draw the parts together.

11 Use a sharp paring chisel and a slicing motion to level the high spots, paying close attention to grain direction.

12 Clean up hard-to-reach spots with a cabinet scraper. Finish up the last bits with sandpaper. It's always used last as stray grit dulls other tools.

A misaligned rail and fitting **is a sign of a misaligned bolt hole. Minor misalignments are to be expected, but this one is too much to fair out.**

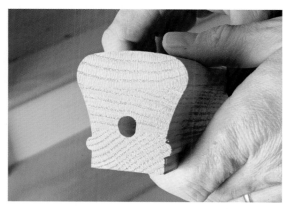

A reamed bolt hole allows proper alignment of railing parts. **Using the 3/8-in. bit, drill out the side of the bolt hole that's opposite the direction the rail needs to move.**

L. J. Smith, one of the leading stair-part suppliers. A nut welded to its center engages the machine threads to turn the bolt.

After checking that the rail bolt is set in the easing so that about half of it is visible in the other piece's 1-in. hole, slide the pieces together to check alignment. Sometimes, things don't fit so well. Rather than proceed with fastening them together and then attempting to recarve the profiles, it's better to disassemble them and ream out the clearance hole until the fit is acceptable.

Now, it may have occurred to you that tightening a nut against the inside of a round hole might move that piece relative to the bolt and hence to the fitting attached to the bolt. You'd be right in thinking that, and some rail bolts today come with an ingenious radiused washer that eliminates that problem. If you aren't so lucky as to get this type of washer, a regular flat washer will also do the trick, as long as you chisel out a flat for the washer to bear on inside the hole.

Joining the parts

When the fit checks out, smear the faces to be joined with 20-minute epoxy. (Don't use 5-minute epoxy here, particularly if the weather is hot: The glue's open time is too short when you're fumbling around starting the nut on the rail bolt.) Then, join the pieces together and slip the washer into place. There isn't a lot of room inside that hole, and getting the nut started can be frustrating. It's all feel and no see. At that, about all the contact you ever have with these nuts is the pad of your index finger. It's helpful to make gravity your ally by holding the assembly so the bolt points up. Sometimes, I spin the nut backward until it engages. When it does, I spin it the other way as far as possible with my finger, then get out a ½-in. box wrench. This is a spot where that L. J. Smith wrench really shines because its narrow waist gives a little more turn before it has to be disengaged and repositioned, but any ½-in. box wrench should do.

Pay attention as you're tightening the nut. Sometimes, the torque will rotate the pieces out of alignment. If that happens, loosen the nut a little if need be, align the parts, and just snug down the nut. Now walk away and give the epoxy time to set up. Come back in a while and finish tightening the nut.

Once the epoxy is fully set, it's time to fair the parts together. I use a variety of tools for this, usually starting with a paring chisel. You've got to pay attention to grain direction because it's possible to start a deepening splinter. Use a shaving motion, and take only a little stock at a time. Other tools I've employed include rasps and files, a block plane, and a cabinet scraper. Sometimes, though, sandpaper alone is enough, and I always finish up by sanding.

Attaching the Newels

As with a post-to-post balustrade, the position of newels in an over-the-top balustrade depends on railing height and baluster location. Install the newels plumb, and be sure to fasten them securely.

Attaching the landing newel

To find the height of the landing newel, you have to know how far it comes past the bottom riser on the upper flight of stairs. I find this on the stairs themselves and then lay out the face of that newel on the full-scale drawing. With the assembled quarter-turn atop the newel, lay the whole shebang on top of the drawing, with the face of the newel on its line. Mark the newel's height on its bottom, then cut to length.

While the newel is laid out on the plywood, I take the opportunity to mark the cut for the quarter-turn's upeasing. This can be done with a pitch block, but it's quicker to use a square to find the tangent point on the upeasing. Square it up from the railing line on the drawing. Lay out the angle of cut in the same way as for the volute, and with the aid of a sacrificial board to

While on-site, **you have to take from the stair the landing newel's location relative to the lowest riser on the upper flight.**

Back to the drawing board. **To find the newel's height and to mark the fitting to length, lay the newel on its line on the drawing.**

More than one way to mark a tangent point. **You can use a square to mark the easing's tangent point directly from the drawing.**

Just like a volute. **Mark the cut line on the easing from a pitch block for that stair.**

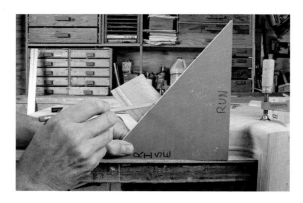

Work around the bulb. Block up the easing to provide clearance for the bulb while cutting. Note the auxiliary fence just visible behind the fitting.

Secure the landing newel **with LedgerLok screws into the framing.**

provide clearance for the quarter-turn's bulb, cut the easing.

Installing the landing newel follows much the same path as its counterpart in chapter 8, with much sawing and chiseling. Once again, LedgerLoks secure the newel.

Attaching the starting newel

The starting newel is straightforward as well. All the hard work is done. The biggest mistake that anyone is likely to make at this point would be to forget to lay out the balusters on the bullnose step before installing the newel. Do that, and the newel will be in the way of the template, which keys off the intersection of the second riser and the finish stringer or skirtboard. This isn't a fatal error, though: Cut out the template to fit around the newel and go.

The volute template comes marked for several baluster widths. The wider the balusters, the fewer are used. Laying out the pattern is a matter of selecting the spacing and, with the template on the step, pushing a pencil or an awl through the center marks into the tread. I like to drill the holes in the tread now as well, before the newel is there and in the way.

Setting the starting newel is a piece of cake. You don't want to slime up the top of the tread with glue squeeze-out, so don't apply glue to the tenon where the act of insertion will squeegee it off into puddles and runnels. Instead, squeeze lots of glue into the mortise, and use a stick to spread it around. Then simply push the newel in. Force may be needed to get the last bit in, and the best way to do that is to drill a dowel hole in a chunk of scrap. Place this block atop the newel and beat the tar out of it with a hammer.

Installing the starting newel

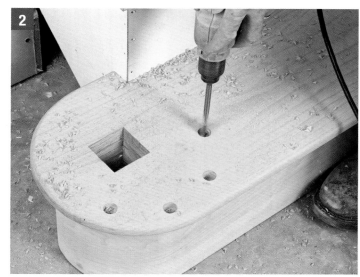

1 Before setting the newel, use the volute template to lay out the balusters. This could be done after setting the newel, but the newel would be in the way.

2 Drilling for the balusters is also easier before the newel is in the way. The baluster holes follow the curve of the volute.

3 Fit the newel to the step. Use lots of glue inside the mortise to secure the newel and avoid squeezing glue onto the tread.

4 That last little bit shouldn't be too easy. Use a beater block with a dowel hole drilled in to complete setting the starting newel. (You want to have this ready before you start—when a newel is jammed partway in and glue is setting is no time to hunt for scrap 2x4s.)

Fitting the Rail

When I first started out, this was the part of over-the-post balustrades that worried me the most. I could imagine making one curved part work, but how was it possible to manage two? As is often the case with stair work, it's just a matter of doing one thing at a time. I make the vertical drop from the quarter-turn longer than the two risers it has to cover and set that in place on top of the landing newel. Using a rail bolt and epoxy, I affix a piece of rail that's a bit long to the volute. That leaves an easing to transition between the vertical drop and that short piece of rail to figure out. And that's not so hard to do.

Mark the rail drop's cut. With the volute level and the drop plumb, mark the upeasing's tangent point on the drop, then cut the drop to length.

Use a square to position the easing. When it's tangent on the drop, put down the easing without moving the square, and mark the cut on the lower rail.

Cut the lower rail to length. A miter saw quickly square-cuts the railing in preparation for bolting the easing to the rail and the volute. A rail bolt and epoxy make the connection between the rail and the easing.

Mark the cut on the top of the easing. After square-cutting the rail drop, put all the parts in place, check them for plumb and level, and mark the easing's cut against the square-cut railing drop.

Measuring for the easing

By temporarily setting the volute and the quarter-turn with it's rail-drop, the easing's location is simple to find. After checking the volute for level and the vertical drop for plumb, find the handrail length by using a square to position the easing along the handrail coming up from the volute. It's right when the easing is aligned with the square and the bottom of the rail, and at the tangent point on the vertical drop. Putting down the easing, mark the handrail to length, remove that assembly from the starting newel, and cut the railing drop.

Now, attach the easing to the volute's handrail with epoxy and a rail bolt, and once again place it on the newel and check for level. With the easing against the vertical drop, mark the tangent point on the drop, remove it from the landing newel, and cut to length. Back in place and plumb and with the volute leveled, simply trace the upper cut on the easing from the bottom of the vertical drop, then take it to the saw and cut.

Before bolting the volute assembly to the quarter-turn assembly, lay the volute upside down on the landing, mark it for balusters using the template, and drill the holes. Drill both the easing and the vertical drop, then fasten them together in place.

Measuring for the upper railing

To find the length of the railing above the quarter-turn, lay a scrap of 1x4 across the tread nosings as a platform for a level. Rest the level on the 1x4 and plumb it up to the bottom of the upeasing; pencil that point onto the level. Moving up to where this railing will die into a trim board on the ceiling, place the level on the 1x4 again and mark the rail height on the

Firmly seat the easing on the saw table **so that its entire edge has contact. The bottom of the easing makes a line of contact against the fence; cut slowly.**

The volute template is finally used on the volute. **Before assembling the volute and the quarter-turn with epoxy and a rail bolt, mark and drill the volute for balusters.**

trim board. (You'd use the same technique if this railing were to meet a newel at the top of the stair.) Measuring between the easing and that point gives the length of the rail. A pitch block gives the cut angle.

Cut the upper rail to length, and it's ready to be drilled for a rail bolt to attach to the easing. At the top, a ½-in. clearance hole from below becomes home to a LedgerLok into the ceiling framing through the trim board.

Measuring for the upper rail

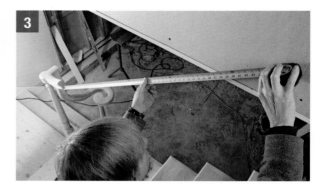

1 The easing establishes the rail height. Use a board to support the level as you pencil the railing height onto it.

2 Mark the upper terminus. Supported by the same board, use the level to transfer the railing height to the ceiling trim board where the railing will die.

3 Stretch a tape to measure the railing length. The cut angle at the rail's top is derived from a pitch block. After it's cut, this rail joins to the easing with a bolt and epoxy.

Installing Balusters

The balusters between the volute and the landing newel are laid out by temporarily installing the rail, then plumbing up from their layout on the treads to the rail. I drill the upper section of rail before it's permanently fixed to the quarter-turn by flipping it upside down and backwards on the stair. That would be too cumbersome for the section between the volute and the quarter-turn, and this is one of the rare times that I drill upward into a rail.

I like to have the balusters below the volute in place when it's time to set it. That's because some of them are farther back on the tread, and you can't get the oblique angle needed to insert the balusters after the railing. I don't want to worry about the balusters moving when the volute is going on, though. If that happens and goes unnoticed, the glue can set with the balusters slightly off the tread or at a weird angle. To avoid these troubles, cut the balusters to length and set them in five-minute epoxy. Make sure the balusters are cut just a bit shorter than the top of the dowel on the starting newel, so that the volute locates on the dowel first when everything goes together.

After the balusters' epoxy has set, mix up some 20-minute epoxy, and coat the dowels atop the newels and the flats around the bases of the dowels with it. Then, lower the assembled railing into place, guiding the balusters into the volute as you go. Once the railing is seated, walk away until the epoxy sets. Once you're confident that nothing will sneak out of place, pin the balusters to the volute.

The rest of the balusters install as on a post-to-post railing. The only other lesson here is setting the balusters into the ceiling trim board at the top of the stair. A laser plumb bob is ideal for transferring the layout up from the treads. You'll have to cut the balusters to length before drilling the holes because you're well into the tapered part of the baluster and the size hole won't be evident until the baluster is cut.

Before packing up my tools, I take a few minutes to look the balustrade over. I look for stray bits of glue at the base of the balusters and check that I've set all the nails. It's a good idea to run your hand over the rail as if you are using it to ascend or descend the stairs, feeling for any spots where the fittings don't transition fairly.

Drilling Fittings for Balusters

When drilling into fittings, the depth of the hole is often limited by the bolts. In fact, rail bolts are a great source of mortality to spade bits. It pays to have spares along. Deal with this annoyance by clipping off the top of the baluster at approximately the angle it hits the rail so that the cut is entirely buried in whatever depth hole in the rail the bolt allows. To fit these balusters, which can't be pushed into a deep hole in the railing, swung over the tread, and their 3/4-in.-long bottom dowel dropped into the hole in the tread, it's often necessary to trim that bottom dowel's length.

Drill the trim board for balusters. Baluster holes in the trim board increase in diameter as the balusters are cut shorter.

Drill this rail in place. After laying them out with a level, drill the baluster holes on the lower level. There are too many appendages on this rail to conveniently flip it upside down on the stairs.

Setting the rail and balusters

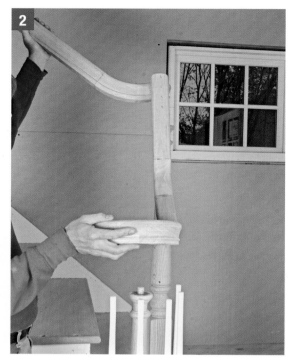

1 Epoxy the volute's balusters to the bull-nose step to ensure that the balusters stay in place while the rail is set. Epoxy's set time depends on temperature: If it's cold, the set is slow; if hot, the set is fast. Check the bond, not the time.

2 Place the rail and fittings as a unit, setting the rail over the epoxy-covered dowels on the newels. The upper end of the rail bolts to the ceiling, plumb above the line of the balusters.

3 Align the volute with the newel dowel first. That way, you're only trying to line up one hole to start. Shorter balusters ensure the volute's placement on the newel before the balusters find their holes.

4 Use finish nails to secure the balusters to the volute. Because the set epoxy reliably holds the balusters to the tread, up-angled nails don't pull the balusters up.

Making Newels and Rails

The next level of craftsmanship above using stock railing parts isn't that big a leap in skills, but it adds a completely new dimension to the look of your work. Railings have always impressed me as the hardest part of stairs, yet when you look at most houses, the balustrades are nearly the same. In fact, as a professional stairbuilder, I loved getting the railing work in a development. I knew all of the houses would use the same parts, and I could buy in bulk to cut my costs. Look through stair catalogs, and you'll see the same patterns of rail, newels, and balusters. Manufacturers call their lines of balustrade parts by different names, but a halfway-close look reveals few real differences.

A further advantage to building your own railing parts is that the material choice isn't limited by some manufacturer's offerings. For the most part, stair parts come in red oak, maple, or beech. What if the house is trimmed in cherry? Or quartersawn white oak?

There's no reason you can't make box newels that work with Colonial, Victorian, or Arts and Crafts designs. You can also build custom railings, even if your stationary shop tools are limited to a tablesaw and a router table (although a shaper would be nice, too). What you can't make with these tools are parts that require lathe work, such as turned newels or balusters. When I've needed custom turned work, I've been able to find local turners who'll do it for me. You also can't make curved rail parts, such as volutes, turnouts, and goosenecks, with a tablesaw and a router. Technically, if you've got a bandsaw and a shaper, you can. For me, that's the limiting factor in custom railing work.

An Arts and Crafts Newel

If you do any casework—cabinets or furniture—making a box newel should come as second nature to you. Essentially, you're making a square tube, then decorating it with a cap, skirtboards, or perhaps moldings and panels. The spare decoration common to the Arts and Crafts style particularly lends itself to box newels.

The strength of a box newel really depends on the strength of the glue joint. A mitered corner exposes more material to glue, making a stronger joint than a butt joint. And unless you're making a stylistic statement, joining the long edges with a miter joint also looks better.

Built-up newels use stock lumber and moldings to achieve a custom look. They were common in Arts and Crafts homes.

The grain on each face of the newel remains consistent on stain-grade work, and miters prevent the butt joints from telegraphing through on painted work. The keys to success are cutting straight and uniform chamfers and using biscuits or splines to align the joint at glue-up.

Getting a strong glue joint from lumber requires surfaces that fit to each other tightly, which means that you have to be able to produce straight joining surfaces. If you buy roughsawn lumber and plane it yourself, you must first joint one face flat. Few of us have a jointer large enough to accommodate stock wider than 6 in. or 8 in., so this can be a problem. I deal with it by buying hardwood from a local supplier that does the face jointing and thickness planing for me.

Straight edges are easily accomplished on a jointer. Rather than joint long boards, which is cumbersome and can eat up a lot of stock, I cut

Box Newels vs. Stock Newels

A box newel is hollow, its sides made from individual lengths of wood. Most are square in section, although I've repaired some Victorian balustrades whose starting newels were large octagonal boxes. Most stock newels are solid, turned affairs. Box newels depend on additional layers such as skirts, moldings, and caps for ornamentation. The fact that box newels are hollow offers at least one advantage: It's easy to run a wire up them. A century ago, when the Victorian and Arts and Crafts eras intersected with electricity, it wasn't uncommon to find box newels with electric lights atop their caps.

Box newels can be made from lumber or plywood. **The square shape is inherently rigid because each side reinforces its neighbors, just like any box beam. The strength of the newel depends on the strength of the glue joint.**

the boards roughly to length. If you don't have a jointer, you can use a long shooting board to cut a straight edge, as shown in chapter 3.

An option to making newels from solid lumber is to use plywood. I do this particularly if I need a smooth substrate for paint or if the newel will have an inset-panel look. Even though plywood is available veneered with almost any species of wood imaginable, it's not my first choice for stain-grade work in a straight-sided newel. The mitered edge is too fragile— it's better to use solid lumber.

Most tablesaws tilt to the right, and most of us keep the fence on the right side of the blade. Having the fence to the left of the blade to cut bevels might seem a bit awkward, but that setup puts the stock above the angled blade, meaning that if the stock rises off the table during the cut, the blade doesn't gouge in as it would if the stock were below.

Making the box

To begin ripping the stock for the miter joint, set up a tablesaw with its blade at 45°. The first cut should be set to leave the stock wider than is needed. What's important is that the cut be dead straight. The straight, jointed edge on the stock runs against the saw fence, which if this were a square cut would be all that was needed to ensure straightness. Bevel cuts are different, though, and if the stock doesn't stay flat on the saw, the width will change.

I prefer to make these bevel cuts with the blade tilted away from the fence. There are two reasons. First, it's safer. If the blade tilts toward the fence and the stock for some reason raises off the table, kickback is likely because the stock is raising into the blade. Second, even if you can control the kickback, the blade just gouged the cut, possibly ruining the stock.

Consistency makes for tight joints. Hold-downs and a featherboard ensure a precise chamfer and an unvarying width. Paper shims create just enough clearance to allow the stock to slide below the hold-downs.

Biscuits spaced about 1 ft. apart **align mitered edges and maintain an even joint.**

Throw-away brushes **spread glue consistently and can be preserved in a plastic bag for reuse.**

Bringing the stock together takes finesse and wiggling. **Start by aligning one biscuit and work your way toward the other end (unless the stock bows out, in which case you align the ends and push in on the middle). Expect a bit of a wrestling match.**

Clamps and a dead-blow hammer close the joint. **The process is incremental—tighten the clamps a little, tap the joint a litle. Eventually, it all comes together.**

Because holding the stock flat to the saw and tight to the fence is so important, I don't try to do it manually. Instead, I rely on shop-made hold-downs that also minimize the risk of kickback. Two hold-downs are spaced from the saw table with scraps of newel stock. Notebook paper shims between the hold-downs and the spacers provide just enough clearance for the newel stock to slide through. A featherboard pushes the stock tight to the fence. For consistency, bevel-cut the first edge of all the newel sides at once. Then, set the saw to bevel-cut them to their final width, and cut the second edges in one group.

Putting together a mitered newel without biscuits to align the sides would be tough. They don't add a lot of strength to the glue joint. In fact, it's not really necessary to glue them. The biscuits are there to ease assembly and to keep the joint lined up. I use a #20 biscuit about every 12 in.

When doing stain-grade work, minimizing glue squeeze-out while avoiding a glue-starved and weak joint is a challenge. I use a foam brush to put a thin coat of glue on each surface before assembly. Working quickly, I slip a biscuit into each slot, and start putting the sides together. Bowed stock is a little tougher, but once you get the biscuits started in the slots on both sides, the stock straightens out. Clamps are always needed, and a dead-blow hammer comes in handy.

Capping a box newel

The newel cap is a big part of the newel's look. It's possible to simply bevel the edges of a chunk of 5/4 or thicker material and nail it atop a box newel. I've been guilty of that more than once while doing production work in tract homes. However, a couple of hours' work can net a far more elegant multipiece newel cap.

Making the cap frame

1 Rabbet the inside edge of the frame stock. One rabbet fits the top of the box newel, and the other provides a perch for the top block. Set the blade ¼ in. high and ¼ in. from the fence to cut the rabbets without resetting the saw.

2 Use a tablesaw to remove most of the stock prior to routing with a panel-raising bit. A feather-board keeps the stock tight to the fence and reduces the chance of kickback. Use push sticks to finish the cut.

3 Two featherboards guide the narrow stock past the panel-raising bit. The vertical feather-board is spaced from the router table's fence by a piece of ½-in. plywood to place downward pressure on the stock's flat and prevent it from tilting inward.

4 To avoid grain tearout, cut miters so that the blade cuts toward the long point.

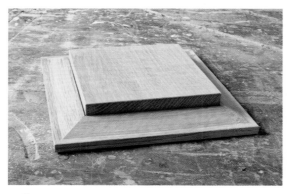

Built-up caps are composed of molded stock put together like a picture frame and a top block that sits inside the frame.

Spring clamps are inexpensive, **leave holes smaller than a finish nail, and exert plenty of pressure to glue up the frame.**

Multipiece caps start with a frame, like a picture frame, that captures the top of the newel on one side and the top block on the other. To begin, cut rabbets into the top and bottom of the frame material. The other edge of the frame stock gets shaped like a raised panel. To do this, first remove most of the waste with a tablesaw and finish the cut with a panel-raising bit in a router table. Now, this sequence has you running some fairly insubstantial stock over some powerful tools. I always run longer stock than is needed and make use of hold-downs and push sticks to keep my fingers out of the blade and bit. At that, I use a speed control on the router table to slow down the large-diameter panel-raising bit, and I make several shallow passes.

All that holds the frame together is glue. After the pieces are mitered to length so that the frame will sit atop the newel, coat both sides of each miter with carpenter's glue, and clamp them together on a flat surface. The miter clamps made by the Collins Tool Co. are inexpensive and ideal for this job. Such

Sanding down the high side **levels an imperfectly matched joint, making it look as if it were perfectly cut and assembled.**

The top block drops right into place. **A couple of drops of glue are all it takes to hold the top block in the frame.**

miters rarely align perfectly—one piece usually ends up just a bit higher than the other. The down-and-dirty solution is to sand them even from the low side, but doing so exposes end grain on the high side that will stain darker. Instead, take the extra time to flush the joint with a sanding block from the high side.

With the frame complete, measure the width of the cap piece and rip sufficient stock to cap all of the newels you're making. Then, cut the cap stock to length, ease its upper edges, and glue it into the frame. Fixing it to the newel takes a couple of drops of glue on the bottom rabbet and setting the cap in place. (I do this on-site because uncapped newels are easier to secure for transport.)

Once on-site, box newels often require blocking for a solid installation. After notching them, I like to glue and screw them into place, often adding the skirts on-site.

A rabbet ensures alignment. **With the block in place, the rabbet that fits the newel is clearly seen around the frame's bottom.**

Installing a box newel

1 Mark newels in place when you can. By holding the newel plumb and marking the landing cut in place, the question of whether the landing is level is avoided and a perfect fit all but guaranteed.

2 If you're proficient with a circular saw, you can make most newel cuts freehand. Clamp the newel to a solid surface, use a sharp blade, and go slow.

3 Since overcutting with a circular saw isn't an option, finish up the cuts with a handsaw.

4 Fill the gaps for a stout attachment. Blocking made from framing lumber is scribed to fit the newel exactly.

5 Trim to the scribe marks, then add blocking to the framing to provide a solid anchor. Glue is cheap, so use lots of it. Drill for the screws to prevent splitting the blocking.

6 Careful work up to this point ensures that the newel will slip snugly around the blocking.

7 Assemble the skirt one piece at a time around the newel, making it slightly oversize (note the shim above the nail gun) to ease installation. Cove molding will trim the top of the skirt.

Locating the Newel

The newel centers on the railing and balusters, but it all starts with the landing tread. The landing tread must bear on the subfloor, and it must extend far enough beyond the wall finishes below, including drywall and skirtboards, to look good. The outside faces of balusters that affix to the landing tread should land inboard of the landing tread's roundover. Once that location is dialed in, you know that the centerline of the balusters, the rail, and the newels is half a baluster thickness farther in.

Doing a quick job-site sketch is a great way to figure out exactly where to set the landing tread and the newel.

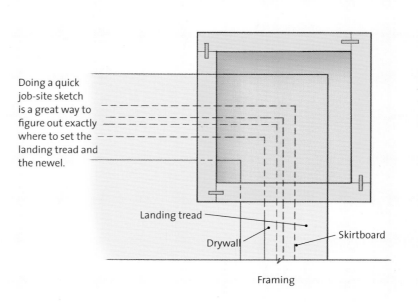

Landing tread

Drywall

Skirtboard

Framing

Built around a basic plywood box, layers of wood and moldings can fit a custom newel to any house's motif.

Tablesawn Spline Joints

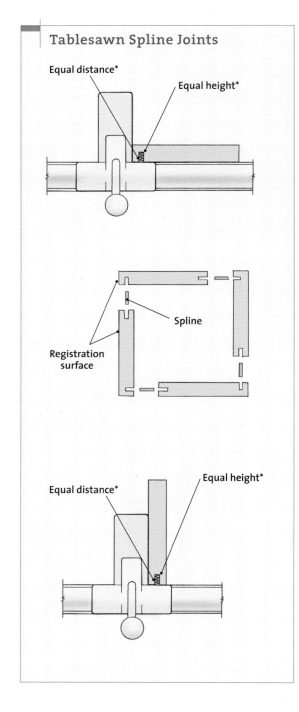

Equal distance*

Equal height*

Spline

Registration surface

Equal distance*

Equal height*

A Victorian-Style Newel

Although the Victorian style and the Arts and Crafts style would seem to be polar opposites, they shared a taste for hefty millwork. The basic box of an Arts and Crafts newel can easily evolve into a more ornate Victorian style by framing the newel with some stiles and rails and adding on a few pieces of molding.

Building the plywood box

Because the newel shown here was destined to be painted and because the stiles would cover its corners, there was no reason to use anything other than plywood for the box (I used ¾-in. utility-grade hardwood plywood).

Mitering the plywood would have been overly fussy, although it would have offered the advantage of a wider glue surface. Instead, I used splined butt joints. The increased surface area offered by the splines makes for a stronger glue joint, and they aligned the edges of the box in a no-nonsense manner.

By making the side pieces ¾ in. (the thickness of the material) narrower than the desired newel width, they could be joined edge to side and all be the same width. The trick to successful and fast spline joints in this case is to be sure to always place the edge or face that would be to the outside of the newel against the saw fence. In effect, that meant remembering to flip the piece end for end between cuts. By doing so, the location of the kerf for the spline didn't have to be perfectly centered or aligned in any special way. The spline itself was ripped from scrap lying on my shop floor.

Dressing up the box

Much like a rail-and-stile cabinet door, the frames applied to the box created an inset panel. The stiles are ripped to a 45° bevel and wrap the corners of the core with long miter joints. Differing from a rail-and-stile door, the grain in all of these pieces runs in the same direction. That's because the height of the bottom rail would be tall enough to create problems with wood movement. Conversely, run as it is, the width of this (and the top) rail is too narrow to move significantly.

The stiles are made from finger-jointed pine, miter-joined to avoid the paint crack that a butt joint would inevitably show. Tacking together a test corner allows you to mark the box and then measure to determine the width of the rails. With that knowledge in hand, rip the rails to width and join them to

Spline joints are quick, strong, and easy. **Two passes per board create kerfs for splines that align and strengthen the joints.**

Glue and spline-join the plywood sides. **Clamps and a few nails will hold the newel together until the glue sets.**

the stiles with glue and pocket screws to create the frames. Leaving the frames long to be trimmed later, glue and tack the four frames to the boxes. Once the glue has set, sand all the surfaces flush and cut the top of the newel to length. Add cove molding inside the frames to complete the look.

Adding a rail-and-stile frame

1 With the saw set to 45°, rip eight stiles the full height of the newel and of a width you find visually pleasing.

2 Holding two stiles together as they'll eventually be assembled, make a mark on each side of the plywood core. The distance between them plus ¹⁄₁₆ in. is the width of the rails.

3 Glue and pocket-screw the rails and stiles on the frame. The pocket screws draw together the joint, acting as clamps for the glue. The bench clamp aligns the faces of the boards, and the screws can be withdrawn once the glue sets (a good plan if the newel will be cut to fit).

4 Glue and set the frame. Use finish nails to secure it to the box until the glue sets, and glue the mitered edges of the abutting frames to their neighbors.

5 Use an orbital sander to level the joint and leave a paint-ready surface.

6 Cut each side with a circular saw and guide to leave the newel's top ready to be capped. Not even a 12-in. miter saw will cut a big newel in one pass.

7 Miter the cove long and snap it in. To get a tight fit, install the short pieces first and then the long, flexible pieces.

A frame of stock chair-rail cap tops the newel. **Other rabbeted moldings, such as window stool, could also be used here.**

Capping the box

I made the cap frame for the newel from stock chair-rail cap, mitered and glued together, and the cap top itself from some 5/4 poplar stock. I wanted a larger raised-panel look than my router could deliver, so I beveled the cap on the tablesaw. A series of cuts around the top finished the raised-panel look, and a sanding block and 100-grit paper made quick work of the saw marks. The bottom of the cap received a rabbet to fit inside the cap frame, and screws from below hold the assembly together. The final step is trimming the bottom of the newel out with a large, 7½ -in. base molding.

Shallow cuts finish the cap bevels. **Be careful not to overcut—it's easier to pare off a little extra material than it is to repair a saw kerf into the bevel.**

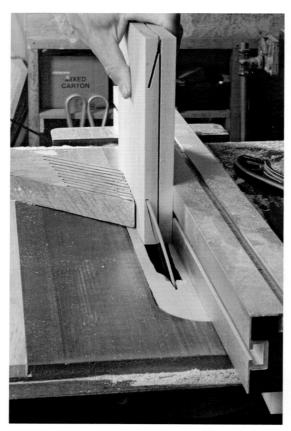

Bevel the cap with a tablesaw. **This cap has a 2-in.-wide bevel of about 15°. The featherboard helps to prevent kickback and ensure consistent cuts.**

Recess the bottom of the cap. **Remove enough material to create a square knob on the cap's bottom that will seat in the chair-rail cap.**

Screws pull the cap's parts together. **The knob on the cap aligns it with the chair-rail frame. Countersunk screws won't keep the cap from sitting solidly on the newel.**

Perfect fit. **The cap isn't affixed yet, allowing access to the inside of the newel to ease attaching the rail.**

Base molding completes the look. **Use a wide base—this one is 7¼ in. —to provide a heavy visual anchor for the newel.**

The railing shown in this section is too large to serve as a code-approved handrail. It works fine as a guardrail, though. A smaller rail was attached to the inside wall to satisfy the building code's handrail requirement.

Making Custom Handrails

Every so often, a job comes along that begs for a custom handrail. The Arts and Crafts stair shown earlier in the chapter was one. Not only did the owners want a heavier rail than is typical, but they also wanted it in quartersawn white oak and they wanted a very specific pattern. Since it was a straight post-to-post railing with no fittings, it was within my abilities.

One complication with this railing was that, in addition to wanting quartersawn stock, the owners also wanted all of the visible faces to be quartersawn. That's not possible with a stacked lamination, the way most railings are assembled. The edges of the various boards would be visible. The solution was to build the rail as a rectangular tube, with the joints between the sides and the top falling within the curve at the top of the handrail. Viewed from above or from either side, the visible wood is quartersawn.

A fully custom balustrade **wouldn't be complete with a manufactured rail. This one was made by profiling a glued lamination.**

To ensure that the rail blanks glued up at a consistent size that could be accurately machined, the top piece of the rail was rabbetted on both sides of its bottom to accept and align the sides. The internal reinforcing blocks were spline-jointed to the sides for accuracy.

A common complaint with stock handrail is that most examples are not only laminated, but also they're made from shorter chunks of wood finger-jointed together. That wouldn't fly in this house. Even though quartersawn stock isn't typically available in long lengths, I was able to make up the blank with only a couple

Assembling a Railing

Often, a simple stack of boards glued together suffices for a railing blank. In this case, the sides and the top all had to show quartersawn grain. Pieces 1 and 3 are quartersawn white oak. Piece 4 is white oak, some quartersawn and some not. Piece 2 is whatever was lying around the shop.

Spline-joining 1 and 2 created accurate rabbets for 3 and 4 to seat in. There was plenty of surface area for a good glue joint and enough reference surfaces so that gluing up a rail blank of a consistent size required only assembly and clamping—no eyeballing one edge to another to ensure alignment.

The center is hollow to ease assembly and to minimize the effect of wood movement. Wood moves more across its plainsawn dimension (perpendicular to the quartersawn dimension) than across its quartersawn dimension. If the assembly of 1 and 2 had met in the middle, the effect would have been that of having plainsawn board glued to a quartersawn board. Over time, the glue joint would have been stressed and would be more likely to fail. By leaving the center hollow, no significant strength was sacrificed, and the differential wood movement was halved.

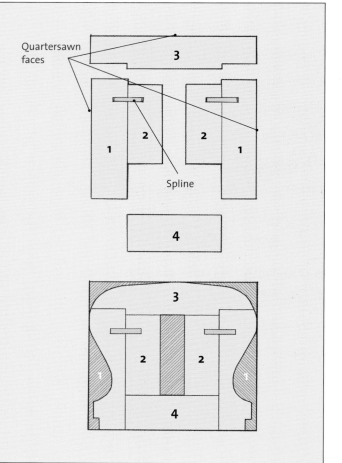

of joints along the length. Lapping the joints by at least 4 ft. with the other parts of the rail eliminated any structural concerns.

Coving on a tablesaw

Shaper cutters are available that will mold just about any handrail you can imagine. As I don't have a shaper, I needed to be creative. Smaller railing bits can be had that work in router tables, and I used one of these to make several of the samples for the owners of this house. None of these samples proved to be suitable for this job.

The biggest problem I faced was that none of the router bits available cut a deep enough hollow to look good on so massive a rail. I'd made large cove moldings before by passing the stock over the tablesaw at an angle to the blade, which is the tack I took on this rail.

The width of a tablesaw-cut hollow or cove is simple to set—it depends on the height of the blade (a higher blade yields a wider cove) and the angle at which the stock is presented to the blade. The closer the angle of presentation is to perpendicular, the wider the cove or hollow. By manipulating these two settings, you can cut a range of depths, widths, and uniform curves.

When coving on a tablesaw, setting the blade at 90° to the table always cuts a curve that's the same on both sides. Setting the blade at a bevel allows you to cut elliptical curves that are shallower on one side than the other. That's what's usually desired in a handrail hollow. This rail was cut with the bevel angle at 21°. That's just a starting point, however. Coming up with the curve that makes you happy takes some trial and error, so it's a good idea to make up half a dozen short rail blanks to experiment on.

These railing samples were all made using a tablesaw and a router table fitted with stock railing bits available from any good supplier.

Oh, the things you can do on a tablesaw. **Stock run obliquely over a blade set square to the table will be molded in an arc of a circle.**

Set the blade to a bevel, **and the hollow will be shaped like a bisected teardrop. The closer to parallel with the blade that the stock is presented, the more exaggerated the teardrop shape.**

Any movement of the rail blank as it passes over the sawblade is a bad thing. The blade can dig in too far and leave you facing a difficult sanding job. You simply can't cove on a tablesaw without setting two parallel fences. I space them apart using a piece of the railing blank. A piece of paper between the setting blank and one of the fences creates enough clearance for the rail to slide easily.

The second key to avoiding movement is to take shallow passes. The first pass or two can be relatively deep, but the amount of the blade contacting the wood increases geometrically with height. The final passes need to be pretty light—¹⁄₁₆ in. or so. Not only do light passes

leave less tearout and finer machine marks, but also the risk of making the blade flex and dig into the blank is reduced.

Blade choice has an effect, too. A blade with more teeth makes a smoother cut, but a blade with fewer teeth is more aggressive. I use a 60-tooth rip blade, and things go well. There's still cleanup, to be sure, but nothing that a sharp scraper doesn't make quick work of. You could also sand out all of the saw marks, but a scraper is much faster.

Finishing up the rail

In this case, my clients wanted a distinct peak to the rail, like the gable of a house. I could easily create that on a tablesaw. The roundover that transitions the flats that form the peak to the railing's side was made with a 1½-in.-radius roundover bit. Like the panel-raising bit discussed earlier, I run this large bit at a reduced speed and take multiple passes. To keep that roundover tangent to the flats, the adjacent flat had to seat on the router table. To stabilize the rail as it passed over the router table for this cut, I used spray adhesive to glue a scrap generated when the flats were ripped to the router table.

Rabbets cut along the bottom edges of the rail completed the machining. Cleaning up the machine marks and easing the transition between the roundover and the cove is handwork. Light passes with a smoothing plane took care of the saw marks on the crest, while a cabinet scraper made pretty quick work of the saw marks on the cove. After I was finished scraping and planing, I eased the transition with a sanding block, and the handrail was done.

Molding a handrail

1 To make the cleanest hollow, take multiple shallow passes and go slow. If the blade starts to flex and vibrate, you're either taking too deep a cut or going too fast.

2 Bevel the top on the tablesaw. A variety of curvy shapes can be molded into the top of a rail with stock router bits. The customer for this rail wanted a distinct gable-roof shape.

3 Set a large roundover bit (1½-in. radius) tangent to the existing molds on the rail. The chamfered strip supporting the rail is scrap affixed to the router table using spray adhesive.

4 A cabinet scraper gets the job done quickly. If you've never used a scraper, you'll be amazed at how much faster than sandpaper it is for some uses.

5 The transition between the rail's hollow and its roundover was too abrupt. A few minutes of block sanding eases the transition nicely.

Building Outdoor Stairs

Building outdoor stairs differs from building indoor stairs in three key areas: material, water management, and measurement. Exterior stairs can have notched stringers or housed stringers. They can have open or closed risers, but remember that the same codes apply to the stairs and rails. The unfortunate fact, however, is that there is little hope that any wooden outdoor stair can last as long as an indoor stair. That said, you can still build outside stairs that will last decades, if you figure out ways to deal with water and sun.

Outside stairs take a beating from the elements. If they aren't soaking up rain or snowmelt, they're baking in the sun. The sun's ultraviolet (UV) rays degrade the lignins in the wood, which are the glue that holds together wood's cellulose fibers. The only way to protect lignins from UV degradation is to coat them with a finish that contains UV blockers. Many clear finishes in particular lack UV blockers and are doomed to fail the moment they're applied.

Another effect of the sun, particularly on wood that's wet from exposure, is differential drying. For example, the shady underside of the tread remains wet, but the side exposed to the sun dries out. Wet wood expands, dry wood shrinks, and the wood cups

Notched-stringer stairs **make a broad and inviting path to a porch.**

Housed-stringer stairs **are a good choice for a large rise because of their overall superior strength.**

and cracks. The resulting cracking, which allows water to penetrate the wood and cracks to collect dirt. Now, wetting a piece of wood causes it to swell, typically closing cracks. But if the crack is filled with dirt, the wood's effort to close the crack simply pushes on the dirt and lengthens the crack. Cracking is a downward spiral.

Find the stair's overall rise. This starts with a guess because to know where to measure from you need to know how far out the stairs come, which can only be known after you've figured out how many treads there are.

Measuring for the Stair

If the ground or sidewalk where the stair will land is close to level, measuring for outdoor stairs is about the same as measuring for interior stairs. When the ground slopes, the location of the stair's landing can be a bit of a moving target, and the more the ground slopes, the faster that target seems to move. The trick is to get the slope of the stair to intersect the slope of the ground.

To start, make an educated guess as to where the stair needs to land. Then, using a long level or a transit to find a level line (a water level would also work), measure the rise from the chosen spot to the top of the deck or porch. Dividing that height by an estimated unit rise of seven gives the number of risers, plus a remainder. If it's a long stair, ignore the

Wood for Outdoor Stairs

Wood rots because various fungi literally eat it. Rot organisms require a trinity: food (the wood), moisture, and warm temperatures. Remove any of these components and rot stops. Temperature is pretty much out of our hands. Paint and other finishes, plus good detailing, can go a long way to keeping wood dry (see the sidebar on p. 208) but aren't entirely reliable. That leaves the wood.

Pressure-treated lumber as well as naturally rot-resistant species such as cedar, ipe, mahogany, teak, and redwood are partial solutions. Other rot-resistant species may be available locally, including black and honey locust, white oak, cypress, osage orange, mulberry, black walnut, and even Douglas fir, if you're in the dry areas of the country. When using a rot-resistant species, avoid sapwood. Sapwood is generally lighter in color and lacks the natural chemicals that make heartwood unappetizing to pests.

Wood and plastic composites such as Trex® are another option that can be used as tread and riser material, if properly supported. Composites come in limited dimensions, generally 5/4 x 4 and 5/4 x 6, and can't span much more than 16 in. If you use them, plan on a lot of stringers.

Pressure-treated lumber

Rot organisms don't eat pressure-treated lumber because the treatment renders it indigestible. Until 2004, wood for residential use was treated with chromated copper arsenate, or CCA. In a voluntary move, the pressure-treating industry dropped this standard brew for one without arsenic. The most common new preservative is ammoniacal copper quaternary, or ACQ.

ACQ-treated lumber is highly corrosive to steel and aluminum. When working with pressure-treated lumber, hot-dipped galvanized or stainless-steel nails, screws, and bolts are required. Stainless steel is best, as its corrosion resistance extends throughout the fastener, not just to the surface layer as with galvanized fasteners. Electro-galvanized fasteners have a thinner zinc coating and are unacceptable.

Another problem with pressure-treated lumber is that the preservative may not fully penetrate the lumber. This is particularly common in the western states, where the species used for pressure-treated lumber is hemlock. Hemlock doesn't accept pressure treatment as easily as the southern yellow pine used elsewhere. The problem arises when the board is cut, exposing untreated wood. The solution is to soak cuts in a copper napthenate solution, such as Osmose® End Cut Wood Preservative.

remainder. If the stair will have less than, say, six risers, and the remainder is greater than 0.5, then round up to the next number of risers. Deduct one from the number of risers to get the number of treads, and multiply the number of treads by the desired unit run to find the overall run. Measure out the overall run level from the structure. Usually, that location is just a refinement of your original guess, and you have to measure the overall rise again and sometimes refigure the overall run, all the while circling in on the sweet spot.

Most of the time, measuring outside stairs isn't that complicated as they typically land pretty close to the structure on relatively level ground. Measuring down from the porch or deck gives a round overall rise, and from that you can derive the overall run. Then, measuring down from a level gives a working overall rise.

Out-of-Level Landings

When a stair lands on an existing out-of-level sidewalk, the approach taken depends on the severity of the slope. If the difference from one side of the stair to the other is less than 1 in., keep the standard rise in the stair's center and split the difference on either side.

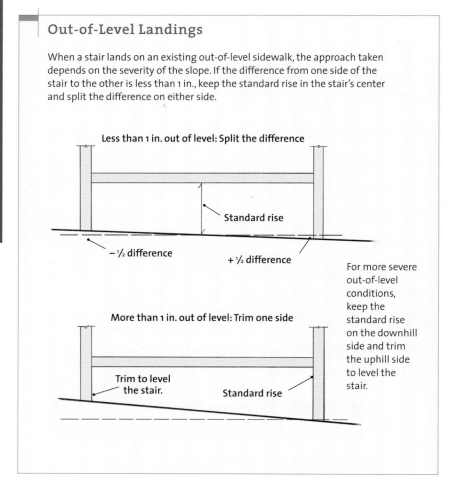

Less than 1 in. out of level: Split the difference

Standard rise

− ½ difference

+ ½ difference

More than 1 in. out of level: Trim one side

Trim to level the stair.

Standard rise

For more severe out-of-level conditions, keep the standard rise on the downhill side and trim the uphill side to level the stair.

Locating the stair's landing pad

Exterior stairs have to land on a pad, something more substantial than dirt. Depending on the job, I've poured concrete pads for stairs to land on, placed concrete blocks, or as here, where a rustic stone walk is planned, set stones. I set the landing pad from the working overall rise and run, then remeasure the rise. This second overall rise, which should be close to the original number, is the one used to lay out the stairs.

To find where to set the landing pad, place a drywall square against the structure aligned with the desired stringer location and then plumb the overall run down from the square.

That point represents the outside corner of the bottom riser, and the landing pad has to extend far enough back from that to support the stringer's bottom level cut. It's important to set the landing pad as level as possible. With a stone pad, level is relative, but do what you can. Each stringer will need a pad, and the quickest approach is to set the two outside pads first. With these pads set level with each other, run a straightedge between them to get the height of the intervening pads.

Three 2x12 stringers are acceptable for this stair because it is short and because its closed risers stiffen the assembly. If I'd been using housed stringers, as in the sidebar on p. 214, two stringers would have been fine. When a notched-stringer stair gets to be six or more risers in height, I add stringers. At six risers, I'd space stringers 2 ft. on center. Around nine risers, I'd reduce the spacing to 16 in., and at 12 risers, I'd go 1 ft. on center. This assumes solid risers and 2x treads. If using 5/4 decking material for treads, I never space stringers more than 16 in. on center. If these rule-of-thumb spacings don't work evenly with the width of the stairs, reduce the spacing; don't increase it.

I'm more conservative in spacing outside stringers than inside stringers for a number of reasons. First, inside stairs are usually supported by walls, whereas outside stairs rarely are. Second, inside stairs aren't susceptible to the ravages of weather. Finally, deck collapses are a fairly regular occurrence, in part because their construction isn't taken as seriously by many as it ought to be but also because when you host a graduation party for a hundred of your closest friends, it tends to be outside on the deck. There is no reason to suspect that deck or porch stairs are any less subject to overloading than the deck or porch itself.

Consider the architecture in sizing the stair. Plan the stringer layout so that any rail works with existing elements such as porch columns.

Keep the stair square to the structure. Using a drywall square in concert with a level is a good way to locate the end of the stairs.

Stairs need to land on some sort of pad. In this case, native stone that will be incorporated with a walk of the same material is the choice.

Keep stairs level side to side. Getting a level line between the landing pads eases building the stairs.

Laying Out and Cutting the Stringers

There are some notable differences between notched stringers outdoors and notched stringers indoors. For one, the material used. As discussed in chapter 2, I don't think 2x12s are acceptable as stringer stock indoors (see the sidebar on p. 25) because better material exists. Outdoors, there's not really a choice. So, I look for flat, relatively knot-free material. Don't worry a lot about there being a crook in the lumber; just lay it out with the crown facing up. Notching the board will relieve the tension on that side, and which way the new internal dynamics of the board will make it move is anyone's guess and not worth worrying about.

Although these are finish stairs, I don't use a miter joint at the intersection of the risers and the stringers. Because of the movement inherent to wood used outdoors, exterior miters just don't hold up well. Unless it's a long-grain to long-grain miter, as you'll see later on the newels, avoid outside miters.

Laying out the stringers

The layout of the basic stringers starts much the same as on inside stairs. On this stair, the porch's band joist serves as a top riser, so the length of the top tread cut has to be shortened by the thickness of one riser (if it wasn't, the tread overhang at the front riser would be short the thickness of the riser).

One big difference between inside and out is that, contrary to the carpenter's paradigm of plumb and level, outside stringers should pitch away from the house to drain water. Too much slope, of course, isn't acceptable on a surface you'll be walking on. Plumbers rely on ¼-in. drop per foot of run for drain lines, and that's what I shoot for with stair treads.

Laying Out a Stringer for Drainage

A pitch of ¼ in. per foot of run encourages water to drain, without feeling too odd underfoot. It also works well with 12-in. treads. You'll have to do the math for shallower treads, or just use ¼ in. and live with a bit more slope.

Start by adding the amount of pitch any given tread would have to the overall rise of the stair. Lay out the treads and risers as you normally would, but go back and lower each rise by the pitch amount. To keep the bottom rise consistent, add the pitch amount to that number. If it's a notched stringer, deduct the tread thickness as usual.

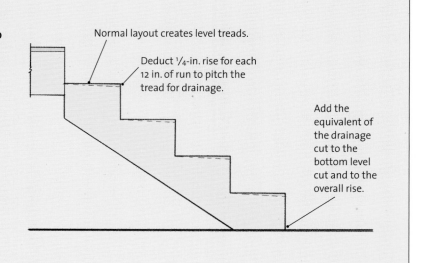

Normal layout creates level treads.

Deduct ¼-in. rise for each 12 in. of run to pitch the tread for drainage.

Add the equivalent of the drainage cut to the bottom level cut and to the overall rise.

Fitting the stringers

1 Lay out the notches with a framing square, then lower the cut line for the treads by ³⁄₁₆ in. to slope the treads to drain.

2 Extend the top of the stringer to come even with the back of the band joist, notching as needed. A vertical 2x4 lagged to the band joist and the back of the stringer adds support.

3 Most outside stringers need to be scribe-fitted. Set the compass to the difference between the level and the tread, then open the compass wider by the desired amount of pitch.

4 Keeping the compass points as plumb as possible by eye, trace the surface of the pad onto the stringer.

5 Cut to the scribe line using a jigsaw. Because the saw vibrates more using an orbital setting, it's harder to follow the line. Use a new blade, and the orbital setting becomes less important.

6 Prime all cuts, and prime again when dry. Keeping water from wicking into the wood matters, particularly in spots touching the ground, like the bottom of a stringer.

To achieve the pitch, first lay out the rise and run as on any stair, but then mark a second tread line below the first. On this stair, I made a mark 3/16 in. down from the front of the tread layouts and extended that line back to the point where the tread and riser meet at the back of the tread. Because the unit run of these stairs is 10 in., a 3/16-in. drop gives roughly a 1/4-in.-per-foot pitch. You do have to remember to cut the lower of the two tread layout lines. In effect, all of the unit rises are 3/16 in. less, so the overall rise of the stair must be raised by that amount so there is no variation at the bottom and top. This is easy—just add 3/16 in. to the bottom plumb cut. Doing so compensates at both the bottom and top risers.

Another bit of strangeness goes on at the top of the stringers. They are extended back and notched to fit under the band joist. The length of the extension is 3 in., give or take, which matches that of the doubled 2x8 band joist. This extension is dropped so that the tops of the stringers, plus the thickness of the tread ends up the unit rise (see previous paragraph) below the porch floor. After the stringers were set, I used LedgerLoks to screw treated 2x4 struts to the back of the band joist and each stringer.

Scribing and cutting the stringers

Because the pads the outer stringers will land on are rough rock, the stringers need to be scribed to fit. Knowing how uneven the surfaces were, I left an extra 1/2 in. when making the level cuts. The 1/2 in. represents the approximate variation in surface height of the stone. With a stringer set so its top is in the proper place, I set a level across one of the tread cuts and set a compass to the deviation between the level and the back of the tread. To this dimension I added 11/16 in. (1/2 in. for the stone's irregularity and 3/16 in. for the pitch) and scribed the bottom of the stringer to fit the stone pad. A jigsaw makes the cut, and the bottom of the stringer is primed.

Attaching the stringers

Attaching stairs to the structure is something few carpenters seem to have the patience or knowledge to do well. I've seem some pretty amazing kludges, from randomly placed toe nails to joist hangers secured with roofing nails. These stairs all had one thing in common: They were being replaced because they were rickety.

The stringers on my stair are supported in several ways. To start with, a cleat is ripped to the width of and screwed to the bottom of the band joist, on the flat. This cleat provides an attachment point for two additional 2x4 cleats to which the stringers are attached with LedgerLoks through their sides. These 2x4s also support the back of the upper tread. Additionally, I install a vertical 2x4 down the back of the band joist and behind each stringer to its bottom. LedgerLoks tie the 2x4 to the framing and the stringers.

Blocking screwed to the band joist provides solid attachment. Notched-stringer stairs rarely provide much direct contact with the structure, so attachment points must be added.

Strong fasteners needed. Attach the cleat that supports the back of the top tread to the porch framing with LedgerLok screws.

A cleat also aligns the stringers. Screw the stringers to the cleat to establish their location and provide initial support. Although the cleat does yeoman's work supporting the tread, alone it's inadequate to support the stringers.

The real support comes from behind. Two by fours lagged to the band joist and to the stringer have the muscle to support a stair for the long term.

Finishes for Outdoor Stairs

Pressure treatment only protects wood from rot. It doesn't protect the wood from UV or water-induced cracking. Finishes do protect wood from UV and water. Pressure-treated lumber, once dried, can be finished like any other wood. With outside stairs, there are steps that you can take as the carpenter to increase the finish's durability and the stair's life expectancy that no one else will have the opportunity to do.

Preparation

Proper application is the key to finish longevity. Wood that's to be painted is supposed to meet the paint manufacturer's specifications for dryness. In the real world, that doesn't happen often: Pressure-treated lumber is usually dripping wet. Buy it as far ahead of schedule as possible, and store the lumber stacked with stickers separating the layers someplace dry (but out of the summer sun to avoid cracks from overly fast drying). These procedures at least dry the surface layers to where they'll hold paint.

Before painting, you should sand all wood with 60- or 80-grit paper to remove milling marks, dirt, and UV-damaged lignin.

Paint

The most important ingredient in paint, the resin, is expensive. Cheap paint is generally so because less resin and more solvent is used. Alkyd resin, or oil-based, primers penetrate and protect raw wood better than acrylic resin, or latex-based, primers. If you can't get alkyd primer, use acrylic primer. Either should contain a mildewcide or a fungicide. Prime all six sides of every board. If you cut the painted board, prime the cut, particularly on end grain. In fact, prime the end grain twice. End grain soaks up water with astounding efficiency. That's its job. End grain is the path that gets water from the ground to the crown of a redwood tree standing a football field in the air.

For a finish coat, use a top-quality acrylic resin latex paint. Alkyd top coats are less flexible than acrylics, and wood outside is going to move. Acrylic paint also allows some moisture transmission, so some gradual drying can take place, while preventing the wood from soaking up gallons of water each time it rains.

Clear finishes

Other finishing options include a clear water-repellent preservative (CWP) containing a fungicide. CWPs are available containing drying oils such as boiled linseed oil or alkyd resin. These are good choices if you intend to glue together the finished parts, but the linseed oil can encourage mildew growth. Be sure there's a mildewcide or fungicide in the finish. If no gluing is foreseen, you can use CWPs with nondrying paraffin oils. As with paints, finishing the end grain is most important. CWPs of either ilk only last a couple of years before recoating is required. Recoating is easy because there's never a film that requires scraping. And they show off woods you might want to look at, such as cedar, redwood, or ipe. Look for CWPs that contain UV blockers.

The other suitable choice is a semitransparent stain. These are chemically similar to CWPs but contain pigments that color the wood and protect it from UV. Semitransparent stains generally require less frequent maintenance and recoating than CWPs, and unlike paint, they don't peel off in sheets.

Clean wood holds paint best. A quick sanding with a 60-grit abrasive removes dirt and UV-damaged wood, leaving a surface that's ready to hold paint for a long time.

Roller priming gets paint on a lot of wood in a little time. Priming should happen right after sanding and in no circumstances more than two weeks later.

Return the backs of the treads to the stringer. Start the roundover 1 in. or so in on the tread's back and continue around to the same location on the far end.

Making water drip. A drip kerf sawn into the bottom of the tread makes water form into beads that drop off, rather than being drawn by capillary action deeper into the stair assembly.

Why a Drip Kerf?

A drip kerf helps to prevent water from getting into the joints of the stair through capillary action. Water's surface tension causes it to stick to surfaces, even to the point of running uphill. For example, water can follow the nosing of a tread around and seep into the joint between the front of the tread and the riser below. Water finds this easier to do on curved surfaces, such as the roundover on treads. Drip kerfs form a sharp edge that makes water more likely to gather into droplets. When droplets form, they can become heavy enough for gravity to overcome surface tension.

Treads and Risers

The first step, so to speak, is cutting the treads to length. They overhang the outer stringers by the same distance as they overhang the risers below—1¼ in. in this case. I used 2x12 pressure-treated southern yellow pine for the treads, and my 12-in. sliding miter saw is just big enough to crosscut them. Next comes rounding over the edges using a ⅜-in.-radius bearing-guided roundover bit. I stop the round-over at the back of the treads about 1 in. from their ends so that it doesn't intersect the risers.

After rounding over the treads, cut a drip kerf in their bottoms. Set the saw to make a ¼-in.-deep cut, and just eyeball its location to be behind the roundover and in front of the risers and stringers.

Risers are even easier to make than treads. Rip to width, cut to length, and scribe the bottom one if need be. Don't forget to prime all exposed surfaces. For the risers on my stair, I chose 4/4 meranti, a rot-resistant hardwood that holds paint well, seems stable, and is quite strong. Like ipe, though, it irritates my sinuses in ways whose description I'll skip. Suffice to say that dust collection is mandatory if you're working this stuff indoors.

The main reason I choose 1x8 meranti over 2x8 pressure-treated pine is that it's nearly impossible to find decent 2x8s (I also think that 2x stock is structural overkill for this application). There seems to be something about the way logs for treated lumber are sawn that makes finding knot- and pitch pocket-free 2x8s rare. That's never a problem with 2x12s, though.

The bottom riser is the first to go on. **Make the attachment with screws, as the risers are structural. They support the fronts of the treads, preventing flex and allowing longer spans between stringers.**

Installing the risers and treads

Starting at the bottom, simply screw the risers into place, using two 2½-in. screws per stringer. (Note that the cuts on the stringer have been primed at this stage.) On painted stairs, I don't worry about the exposed screw heads. They blend in with a coat of paint, and in truth, people aren't surprised by exposed fasteners in exterior work. If hidden fasteners were expected on the stairs, I'd drill oversize holes and plug them.

After the risers are all installed, turn to the treads. Starting at the bottom, they receive three 2½-in. screws per stringer. Confidence lets you get away with a lot in life, and one key to leaving fasteners exposed is to make it look deliberate. I use a square to evenly space and align the screws in the stringers, and I space screws into the risers at an even interval of about 1 ft. Just as important is to screw through the back of the risers into the treads. Screwing the treads, risers, and stringers together creates a stair that acts as a unit. No single member is ever loaded without support from the other members, and that makes for a sturdy, safe stair.

Even spacing makes exposed fasteners look deliberate. **Unless you're plugging each screw hole, exposed fasteners are de rigueur. Paying attention to spacing improves the look.**

Screw the risers to the back of the treads. **Like the risers in front of the treads, those in the back also play a role in reducing the treads' deflection under load.**

Housed-Stringer Stairs

The stairs I used on my porch are notched-stringer stairs, but housed-stringer stairs can also work well outdoors. The jig used to rout the stringers differs from that shown in chapter 4 (see pp. 61–64) only in that the tread cutout is larger to create a mortise for 2x tread stock, as opposed to 5/4 stock. The assembly proceeds in the same way, treads first followed by risers. A pocket-hole jig is used to create the screw holes in the backs of the risers, but instead of affixing the risers to the treads with regular pocket screws, I use decking screws.

Attaching a housed-stringer stair is in some ways easier than attaching a notched-stringer stair. That's because the tops of the housed stringers make full contact with the joists. Initially, screws through the top riser into the deck or porch framing will hold the stair until you can run some long lag bolts or LedgerLoks through the back of the framing into the stringers.

Use a plywood jig and a bearing-guided bit to rout mortises in a 2x10 stringer.

The first step in the assembly is to wedge the treads into the mortises.

Stainless-steel deck screws join the riser to the back of the tread.

A pocket-screw jig and a proprietary bit make angled screw holes to join the top of the risers to the treads.

Details for Balustrades

Outdoor balustrades are subject to damage from the weather. Joints are often where moisture troubles begin. In fact, whenever I've torn down an outdoor structure, I've almost always found that the joints were where the fatal rot began. Water that enters a joint isn't exposed to the sun and airflow that can dry it.

There are several approaches to avoid trouble at joints. One is to figure out ways to minimize the contact area between the boards, but there are limited spots where this can work. Another is to encourage drainage by sloping the wood. For example, the top of a railing or a newel should never be flat. Upper layers that overhang lower components help keep them dry, and sometimes, you can produce joinery that's tight enough to keep water out.

Now, you might say that I'm giving contradictory advice here, but which advice you take

An open bottom rail encourages water to drain. Additionally, the bottoms of the balusters are exposed to the air and not sitting in a puddle as can happen with a solid bottom rail.

Use the top railing like a roof. Overhanging construction protects the end grain of the balusters' tops. The side trim makes the rail look heavier and keeps water farther away from the tops of the balusters.

Railing Details for Drainage

Since the top railing protects the top of the balusters from rain, a dado and fillet system works well here (option 1). Capturing the bottom of the balusters between two boards, whose tops are beveled to shed water, provides a secure connection that drains (option 2).

If you've got a tenoning jig for a tablesaw, another option is to make two opposing angle cuts to the bottom of the balusters, resulting in a shark's-mouth look (option 3). Rip a single bottom board to the corresponding angle, and you've got a bottom rail that sheds water combined with a simple and secure baluster connection.

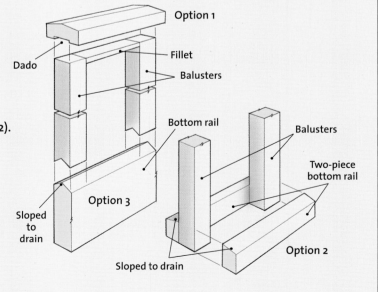

is dictated by the situation. So, with a housed-stringer stair, for example, it makes sense to make the joints between the risers and treads and the stringers as tight as possible, leaving no room for water to get in. Caulk or sealants applied externally to a joint are as likely to trap water in as to keep it out, but completely coating all mating sides of the pieces to be joined with exterior glue (which can be Titebond II or III, polyurethane, or construction adhesive) seals out water.

Outdoor Newels

An outside newel is often as simple as a chunk of 4x4 with its top beveled in an attempt to shed water. While beveling the top beats leaving it flat, you're still exposing the end grain and that post will have a shorter life span. A better solution is to add one of the commercially available copper caps, which do a great job of shedding water.

A 4x4 newel has quite a few limitations. Research has shown that notching a 4x4 to fit the framing pretty much ensures that it can't take the required 200-lb. sideload without splitting. To avoid this problem, and the rather industrial look of 4x4 posts, I prefer to make built-up hollow newels much as those described in chapter 10. As with the risers, meranti is an excellent choice, but any of the rot-resistant hardwoods work. The only disadvantage of a built-up newel is that it's more expensive to make. That's balanced by better looks and higher strength.

Making an outdoor newel

Since I was going to the trouble of making the newels for this stair on my house, I gave some thought to their width. The porch posts are 6x6s, and the newels had to be subordinate in size to them. I didn't want to make the newels the same size as a 4x4, so I split the difference between those two timbers and made the newels 4¼ in. square. They complement rather than compete with the columns.

Starting with four 9-ft.-long 1x6s, first cut them in half for convenience, then straighten one edge on a jointer. Next, rip the opposite edge on all the stock at a 45° angle. Following that, set the fence to the desired width of 4½ in. and rip the second side at 45°.

Prime the front and back of each piece but not the beveled edges. The edges have to make a sound glue joint, and paint would interfere. I used a biscuit joiner to slot the edges at about 1-ft. intervals and started gluing and putting together the newel's sides. Use clamps and a dead-blow hammer to close up the joints.

To avoid miter joints in the newel cap, as were done on the interior newels in chapter 10, I glued up a double thickness of 4/4 meranti that was a little longer than the combined width of two newel caps. After ripping the

Bevel the caps using a sled. **Screwed from the inside to a plywood box sled, the stock for the cap receives its second bevel.**

Making a newel

1 | Bevel the edges of the newel sides on a tablesaw. Hold-downs ensure a consistent angled cut.

2 | Too much glue is not enough. Biscuits and exterior-rated glue make a mess but survive the elements well. The water-resistant glue not only joins the wood but also helps seal out water.

3 | Use a dead-blow hammer, followed by clamps, to coax together the newel.

stock to width, I affixed it to a sled—essentially an open plywood box—with 1¼ -in. wood screws and beveled it to a point on the table-saw. After penetrating the ¾-in. plywood of the box, only ½ in. of the screws went into the caps, not enough to encounter the sawblade. After the second pass left the stock shaped like a gable roof, I removed it from the sled and cut the two caps to length. I screwed both caps to the sled again, and in two more passes they looked like the hip roof of an American four-square house.

Two for the time of one. **Cut to length and affixed to the sled, both caps are finish-beveled at the same time.**

Avoiding exposed fasteners in the cap. **A block screwed to the bottom of the primed cap slips into the newel.**

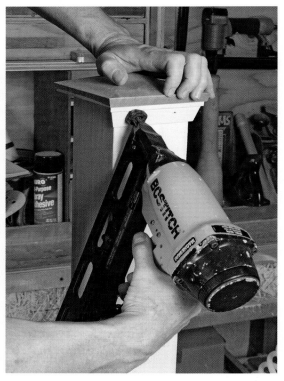

Trim nails serve two purposes. **Long galvanized finish nails penetrate the molding and newel into the cap's block, securing the cap.**

A bit of planing and sanding cleaned up the caps, and their edges were nipped off on the tablesaw to an angle complementary to the top bevel. I screwed square 5/4 blocks dimensioned to fit just inside the newels to the bottoms of the caps, primed the whole assembly, and then set it in place.

Although I do my best to avoid miter joints in the outside, once moldings come into the picture, miter joints are a fact of life. So, I trimmed the undersides of the newel caps with a band molding I made on the tablesaw from meranti scraps. The finish nails that hold on the band molding are 2 in. long—enough to penetrate through the sides of the newels and into the block that's fixed to the bottom of the cap. Those nails are what hold the caps in place, avoiding any fastener holes in the caps themselves.

Attaching the newel

There isn't a lot to attach these newels to— only about 6 in. in height of stringer and riser. To beef up this attachment, add some 2x blocking. Cut back the tread so that when the blocking is installed, the side of the newel slips tightly between the blocking and the tread. Glue the blocking to the stair and secure with several 3½-in. screws and a 4½-in. Ledger-Lok that's run into the tread from the front of the blocking.

Notch the corner of the newel to fit the stair, prime the cuts, and drop the newel in place over the blocking. The newel should be trimmed so that its bottom ends up at least ¼ in. above the pad or sidewalk. Even though this end grain is primed, keeping it out of a puddle is a good plan.

Attaching the newel

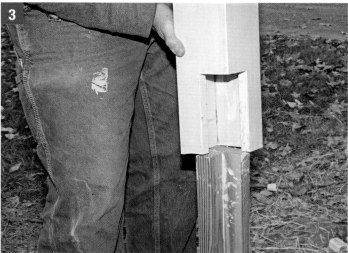

1 | Use a jigsaw to notch the bottom tread even to the faces of the riser and stringer.

2 | For the newel to be plumb, make the blocking plumb. Leave space between the blocking and the notched tread for the newel.

3 | Notched to fit to the bottom step, the newel slips over the blocking. Plane a slight bevel on the blocking's outer corner to provide clearance for hardened glue inside the newel.

4 | Use screws to fasten the newel to the blocking through ½-in. holes that will be plugged.

Although exposed fasteners don't bother me on the stairs, they do for some reason on the newels and the rail. Call me inconsistent. So, I set the screws holding the newels to the blocking in ½-in. holes, which are later plugged.

Because the porch and stair here are less than 30 in. above grade, no guardrail was required but a handrail was. This one was custom, left over from another job, and I shamelessly pressed it into service. As explained in chapter 8, mark the pitch of the stairs on the posts with a level, measure up from that, and mark the rail height on the newel and on the porch column.

Lay the uncut rail across clamps affixed to the newel and the porch column. The clamps are easily rotated up or down to tweak the rail into place. Scribe the cut line onto the rail, cut it to length, prime the cuts, and screw the rail to the posts, bottom first. I epoxy the plugs into place, as that's about the most waterproof glue available. Using the five-minute variety gives me the nearly instant gratification of paring the plug flush.

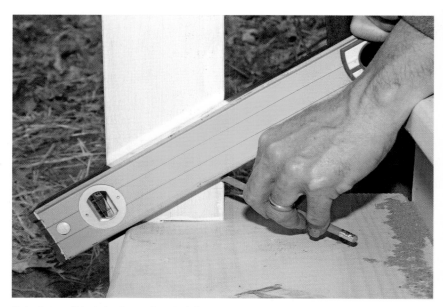

Calculate rail height from the pitch line. **Lay a level across the treads to mark the stair pitch on the newel.**

Mark the top of the rail on the newel. **Measuring up the desired rail height from the pitch line positions the top of the rail.**

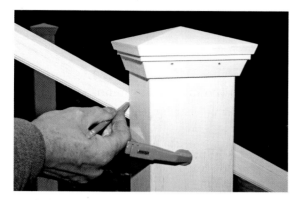

Mark the rail length in place. A clamp closed on the newel supports the rail and can be pivoted up or down to make fine adjustments to rail height.

Drill a pilot hole in the plug hole. The ½-in. plug hole is square to the rail bottom, but the pilot hole for the screw angles inward. That allows the pilot hole to be farther up the rail, leaving plenty of supporting wood to ensure a stout attachment.

Trim the plugs flush. Made from the same material as the rail and epoxied into place, face-grain plugs disappear when painted and keep water out.

The last step

This being the end of a book that I hope has given you hours of enjoyable learning, I'd like to borrow a little more of your time to sum things up.

There's a lot of specific advice in the previous pages, but it shouldn't be taken as the last word on stairbuilding. Over the years, I've made it a point to speak with other stairbuilders. While there's always some commonality to our methods, I've never met anyone who worked exactly as I do. That's not to say their stairs aren't the equal of mine, only that they build stairs in ways that suit them. I've borrowed some of their techniques, and maybe they're using some of mine. I hope so. Since we lack the formal apprenticeship programs of the past, the future of our trade is in the sharing of information over coffee in a dusty, unfinished house, over beer on a pickup's tailgate, at trade shows, online, and in the pages of magazines and books. As you figure out your own ways, please, don't be stingy in sharing them with others.

Appendix

Good tools make craftsmanship possible, although they don't guarantee it. The next few pages contain photos and descriptions of some of my favorite tools. You'll notice that they aren't pristine, which is because they're my daily users. A couple of them have been with me for 20 years or more. There's a lesson in that—cheap tools rarely pay. Buy quality, because just as good tools enable craftsmanship, poor tools don't.

Several railing manufacturers, notably L.J. Smith (740-269-2221, www.ljsmith.com) and Crown Heritage (800-745-5931, www.crownheritage.com), sell several dedicated railing installation tools. Some of them look pretty slick, but I haven't used them and so can't offer up an informed opinion. Most of the tools in my quiver serve me as well for trimming houses, and the need for specialty railing tools just never seems urgent.

Tools

Clamps

A carpenter can never have enough clamps. I rarely buy clamps new, but it's equally rare for me to pass them up at tag sales. Clam Clamps are the heaviest duty miter clamps I've used. I get mine from Chestnut Tool and Chowder Co., 1-800-966-4837, www.miterclamp.com. Spring miter clamps and their accompanying spreader pliers, available from Collins Tool Co., 888-838-8988, www.collinstool.com, pinch together smaller moldings.

Routers and router bits

Three routers get me through most stair jobs. The Mustang (Bosch Tool Corporation, 877-267-2499, www.boschtools.com) laminate trimmer is light and wieldy, great for cutting small mortises. The Porter Cable 690 (Porter-Cable Corporation, 888-848-5175, www.portercable.com) is a medium-duty machine that sees duty cutting profiles both hand held and in a router table. And I have a big Bosch plunge router (a 2½-hp model that's no longer made) that's been mortising stairs for me since 1988.

A variety of router bits produce a variety of molded shapes, from tread nosings to the coves used on custom newels to railing profiles.

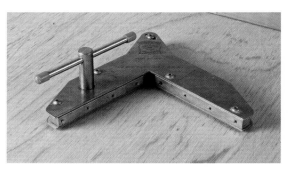

Clam Clamps exert tremendous pressure while pulling mitered parts together evenly.

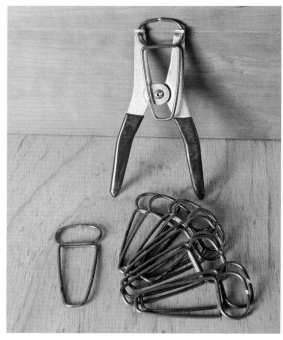

Spring miter clamps leave holes smaller than most finish nails.

A CMT 811.690.11B pattern **routing bit** (C.M.T. Utensili S.p.A., 888-268-2487, www.cmtusa.com) is used in conjunction with a plunge router and plywood jig to mortise housed stringers.

The bits used to produce a variety of molded shapes all have ¹/₂-in. shanks. Except for the lightest work, ¹/₄-in. shanks aren't acceptable.

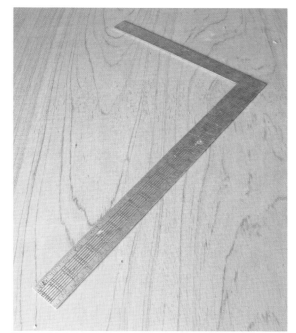

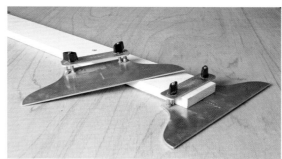

Tread templates (available from Collins Tool Co.) combine with a scrap of 1x4 to accurately measure treads and risers on site-built stairs.

Twelve-inch dividers (Lee Valley Tools, Ltd., 1-800-267-8735, www.leevalley.com) are indispensable for laying out winder stairs.

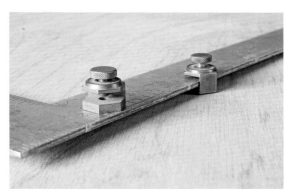

A framing square (top) and a set of stair gauges are the stairbuilder's basic layout tools.

A Starrett protractor is a grand tool for measuring angles. It has two scales: one provides the measurement of the angle in degrees and the other displays the miter angle.

Layout tools

The stairbuilder's most basic layout tool is the framing square (The Stanley Works, www.stanleytools.com). My favorite one is aluminum, as opposed to steel, because it's lighter and it doesn't get unbearably hot if left out in the sun. Stair gauges are readily available at hardware stores, and they work with the framing square to ensure consistent layout. Other layout tools I couldn't live without include a combination square, tread templates, 12-in. dividers, and a protractor.

A laser plumb bob speeds initial layout and could be used to transfer the location of balusters from the treads to the railing (DeWalt, 800-433-9258, www.dewalt.com).

The VersaTool **is a dedicated stair tool used for driving rail bolts.**

Double-end lag screws **attach dowel-less balusters to treads. The aluminum driver shown is a generic version.**

Rail joining

L. J. Smith's VersaTool is one of the few dedicated stair tools I own, and it was given to me by a lumberyard where I bought rail parts. The fact that it was free in no way diminishes its value. The VersaTool drives rail bolts readily, and its hourglass shape reduces the tedium of tightening nuts inside a rail. It's over-rated as a layout tool, though.

Rail bolts now come with a hemi-cylindrical washer that doesn't push rail parts out of alignment. That's great, but the star nuts that still come with the bolts are useless. Buy and use hex nuts. I now use LedgerLok screws (OMG, Inc., 800-518-3569, www.fastenmaster.com) in place of lags to secure newels and rails. Double-end lag screws quickly attach dowel-less balusters to treads. Drivers for double-ended lags are made from aluminum into which the lag threads and is driven into the tread.

Safety

I first discovered the comfort of Peltor Optime 105 (Aearo Corporation, www.peltor.com) ear muffs as an assistant editor at *Fine Homebuilding* magazine. Copying more experienced staffers, I wore them at my desk to cancel out office noise and foster concentration in my cubicle. They work just as well for cutting out noise in the shop or on the job. Safety glasses are also an essential item.

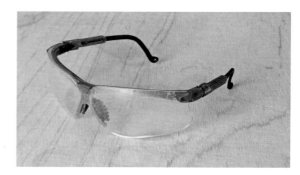

Peltor Optime 105 ear muffs **are a particularly comfortable and effective model.**

Ever visited the doc **to have a foreign object removed from your cornea? I have. Wear ANSI Z87 rated safety glasses. I like these by Uvex (Bacou-Dalloz Corp., 800-343-3411, www.uvex.com)**

Dead-blow hammers **don't leave dents. The color's a grabber too.**

Cabinet scrapers **are cheap, faster than hand sanding, and easy to resharpen.**

Hand tools

Dead-blow hammers are hollow plastic affairs partially filled with shot. Inertia leaves the shot at the back of the hammer head when you start the swing. Because the shot is float-ing inside the hollow head, it's only the mass of the plastic that hits the wood you're beating in the first micro-second, and that's not heavy enough to cause dents. With the plastic ham-mer head stopped against the wood, the shot catches up and delivers the blow, cushioned by the plastic. These things are magical, and the color is styling.

Two styles of chisel cover most stairbuilding bases. The paring chisel is ground at a shallow angle, about 25°, which allows it to be honed like a razor. Paring chisels are pushed by hand, never struck, and they're great for cleaning up that last little bit. Bench chisels are shorter, tougher, ground at a steeper angle of around 35°, and will take being hit with a hammer to create mortises.

Older Stanley No. 4 smooth planes can be had for $20 or so at flea markets. Sharpened up, they flatten glue-joined boards faster and more quietly than any power tool. Cabinet scrapers can be had through any woodwork-ing catalog. They're cheap, faster in many instances than hand sanding, and easily resharpened.

Power hand tools and jigs

Impact drivers use a sort of internal rotating hammer and anvil to drive screws when the going gets tough. The older 12-volt model below (Makita U.S.A. Inc., 714-522-8088, www.makita.com) combines light weight and power in such a way that I can't imagine needing a higher voltage tool. Two downsides to impact drivers: Their hex chuck limits their ability to drill holes, and they're fairly noisy.

Impact drivers **are used to drive screws when the going gets tough.**

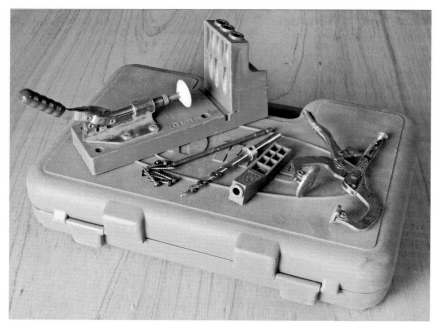

Pocket hole jigs guide a special bit to create an oblique pilot hole.

Pocket hole jigs (Kreg Tool Company, 800-447-8638, www.kregtool.com) guide a special bit to create an oblique pilot hole. Screws driven through these holes pull abutting stock down tight, making for fast and easily aligned joints.

Shop-made tools

The most important shop-made tools of a stair-builder are the mortise jigs. Made from void-free plywood glued up to a thickness of 1 in., these work in concert with a pattern routing bit. Shooting boards are made from Masonite and plywood to guide circular saws, providing straight, splinter-free cuts.

The lower, longest mortise jig is for routing wide treads in winder stairs. The one on the left is most typical, for use with 5/4 treads and 4/4 risers. The right hand jig is for 2x treads.

Shooting boards are made from Masonite and plywood to guide circular saws, providing straight, splinter-free cuts.

Codes

The most commonly accepted building code in the land is the 2006 International Residential Code. It's got a lot to say about stairs, and what follows covers only the highlights. Further complicating matters is the fact that jurisdictions are free to amend the IRC as they see fit, and many do. Before building stairs or installing railings, it's smart to check with the local inspector.

Headroom

The minimum headroom above a stair should be at least 80 in. measured plumb up from the line of the tread nosings. The minimum headroom over a landing is also 80 in., measured finished floor to finished ceiling.

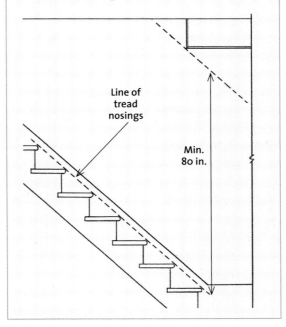

Minimum Width

Stairways must be a minimum of 36 in. wide, and landings must be at least as wide as the stair. Landings at least 36 in. square are required if the stair climbs more than 12 ft., or where a door swings over a stair.

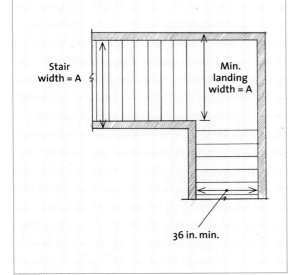

Stair Geometry

In addition to specifying the ranges of unit rise, unit run, and overhangs, the IRC also requires consistency. In each of the latter examples, the greatest within a stair should not exceed the least by more than 3/8 in. No overhang is required when the tread width exceeds 11 in. The maximum space between open risers on stairs with a total rise exceeding 30 in. is 4 in.

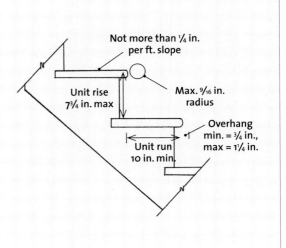

Winder Geometry

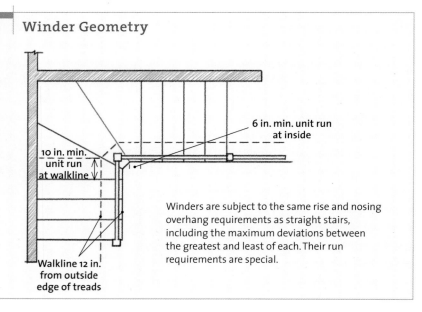

6 in. min. unit run
at inside

10 in. min.
unit run
at walkline

Walkline 12 in.
from outside
edge of treads

Winders are subject to the same rise and nosing
overhang requirements as straight stairs,
including the maximum deviations between
the greatest and least of each. Their run
requirements are special.

Continuous Handrails

Every flight of stairs with four or more risers must
have a continuous rail that, at the least, begins
plumb above the lowest tread nosing and extends
to plumb above the highest tread nosing. Continuous rails may be interrupted by newels at turns.

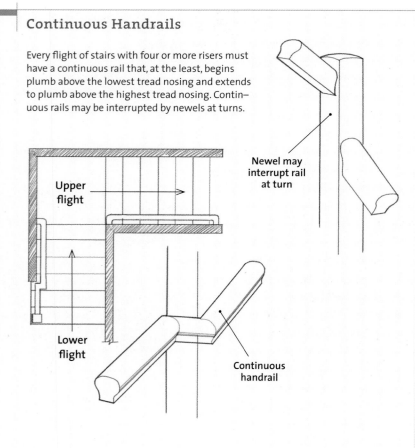

Newel may
interrupt rail
at turn

Upper
flight

Lower
flight

Continuous
handrail

Stairway Minimums

Handrails are required on at least one
side of any stair that has four or more
risers. Minimum stairway width, handrail
projections into that minimum, and when
handrails are required are all covered by
the IRC.

Min. width 36 in.

Max.
projection
4-1/2 in.

Return rail
to wall

Min. space
1-1/2 in.

Two rails require
min. 27-in. walk space

One rail requires
min. 31-1/2-in. walk space

Min. 34 in., max 38 in.
above tread nosings

Handrail Heights and Spaces

Perhaps the most confusing part of the code is the distinction between a handrail and a guardrail. Guardrails are required whenever a stair or balcony exceeds 30 in. high. On balconies, guardrails must be at least 36 in. high. On stairs, they must be at least 34 in. high. Guardrails can also serve as handrails, as long as their height doesn't exceed 38 in. above the tread nosings. Minimum baluster spacing is determined by the size sphere that must be excluded from the space; it varies with location. All guardrails must withstand a 200-lb. sideload.

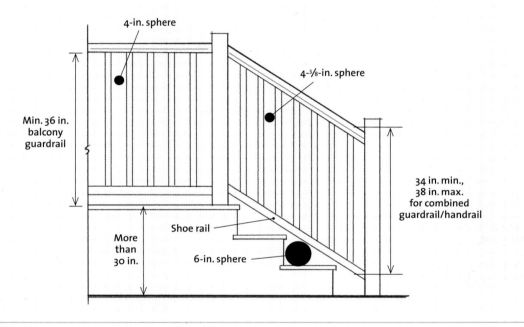

4-in. sphere

4-³⁄₈-in. sphere

Min. 36 in. balcony guardrail

More than 30 in.

Shoe rail

6-in. sphere

34 in. min., 38 in. max. for combined guardrail/handrail

Handrail Cross Section

Handrails must be graspable—neither too large nor too small. Shown here are common parameters, but others may be permitted by local authorities.

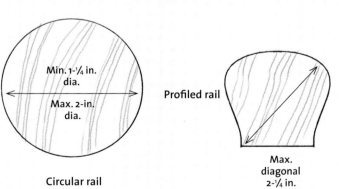

Min. 1-¹⁄₄ in. dia.

Max. 2-in. dia.

Circular rail

Profiled rail

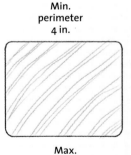

Max. diagonal 2-¹⁄₄ in.

Min. perimeter 4 in.

Max. perimeter 6-¹⁄₄ in.

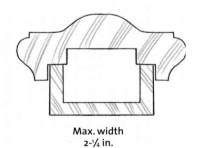

Max. width 2-¹⁄₄ in.

Index